Lecture Notes in Computer Sc
Edited by G. Goos, J. Hartmanis, and J. v.

T0237585

Springer

Berlin
Heidelberg
New York
Barcelona
Hong Kong
London
Milan
Paris
Tokyo

Gholamreza B. Khosrovshahi   Ali Shokoufandeh
Amin Shokrollahi (Eds.)

Theoretical Aspects of Computer Science

Advanced Lectures

Springer

Series Editors

Gerhard Goos, Karlsruhe University, Germany
Juris Hartmanis, Cornell University, NY, USA
Jan van Leeuwen, Utrecht University, The Netherlands

Volume Editors

Gholamreza B. Khosrovshahi
Institute for Studies in Theoretical Physics and Mathematics (IPM)
Tehran, Iran
E-mail: rezagbk@ipm.ir

Ali Shokoufandeh
Drexel University, Dept. of Mathematics & Computer Science
Philadelphia, PA 19104, USA
E-mail: ashokouf@mcs.drexel.edu

Amin Shokrollahi
Digital Fountain, Inc.
39151 Civic Center Drive, Fremont, CA 94538, USA
E-mail: amin@digitalfountain.com

Cataloging-in-Publication Data applied for

Die Deutsche Bibliothek - CIP-Einheitsaufnahme

Theoretical aspects of computer science : advanced lectures / Gholamrez
B. Khosrovshahi ... (ed.). - Berlin ; Heidelberg ; New York ; Barcelona ;
Hong Kong ; London ; Milan ; Paris ; Tokyo : Springer, 2002
  (Lecture notes in computer science ; 2292)
  ISBN 3-540-43328-7

CR Subject Classification (1998): G.2, F.2, F.1, C.2, E.1, I.3.5

ISSN 0302-9743
ISBN 3-540-43328-7 Springer-Verlag Berlin Heidelberg New York

Springer-Verlag Berlin Heidelberg New York
a member of BertelsmannSpringer Science+Business Media GmbH

http://www.springer.de

© Springer-Verlag Berlin Heidelberg 2002
Printed in Germany

Typesetting: Camera-ready by author, data conversion by PTP-Berlin, Stefan Sossna
Printed on acid-free paper   SPIN: 10846385   06/3142   5 4 3 2 1 0

Preface

The First Summer School on Theoretical Aspects on Computer Science was held at the Institute for Studies in Theoretical Physics and Mathematics (IPM) from July 3-10, 2000.

The idea of organizing a summer school was originated during a conversation between the first and the second editor of this volume in the winter of 2000 at IPM in Tehran. The decision was formalized when the third editor visited IPM later in the winter.

The main idea behind the Summer School was to introduce to the students in Iran some of the highly innovative and promising research in Theoretical Computer Science, and, at the same time, to start a flow of ideas between top researchers in the field and the Iranian scientists. For this reason, the talks were designed to be of a tutorial style.

The organizers decided to concentrate on a few important areas of research in modern Theoretical Computer Science. To attract students to this important area, they also decided to include talks on applications of methods from Theoretical Computer Science to other areas.

This volume contains the written contributions to the program of this Summer School. It is intended for students and researchers who want to gain acquaintance with certain topics in Theoretical Computer Science. The papers on quantum computation, approximation algorithms, self-testing/correcting, algebraic modeling of data, and the Regularity Lemma correspond to the former category, while connections between multiple access communication and combinatorial designs, graph-theoretical methods in computer vision, and low-density parity-check codes belong to the latter category.

Each contribution to this volume was presented in a two hour session, leaving ample time for questions and discussions. The tutorials were followed by a number of specialized talks touching on current research problems in the area. Parts of the talks have been incorporated into the material of the contributions.

January 2002

G.B. Khosrovshahi
A. Shokoufandeh
A. Shokrollahi

Organization

The first International Summer School on Theoretical Aspects of Computer Science was organized by the Institute for Studies in Theoretical Physics and Mathematics (IPM) in Tehran, Iran.

Program Committee

Conference Chair:

G. B. Khosrovshahi, IPM and the University of Tehran, Iran
A. Shokoufandeh, Drexel University, USA
A. Shokrollahi, Digital Fountain, USA

Local Arrangements:

M. Zamani

Sponsoring Institutions

School of Mathematics, Institute for Studies in Theoretical Physics and Mathematics (IPM), Tehran, Iran

World Mathematical Year (WMY) 2000 National Commission

Service de Coopération et d' Action Culturelle, Ambassade de France en Iran

Neda Rayaneh Institute (NRI)
http://www.neda.net.ir

Acknowledgements

We are grateful to Professor M. J. Larijani, the director of IPM, who has been a great advocate for the expansion of Theoretical Computer Science at the Institute, and helped us secure major funding from IPM. We would also like to express our thanks to Professor S. Shahshahani, the Deputy Director of IPM, who was most helpful in soliciting financial support from the World Mathematical Year 2000 National Commission. The Service de Coopération et d' Action Culturelle of the French embassy in Iran kindly provided travel funds for our invited speakers from France. We thank the Service Culturelle for their generous support. Finally, we would like to express our gratitude to M. Zamani, the executive officer of the School of Mathematics, who single-handedly assumed almost all the responsibilities for the local arrangements of the summer school, and was crucial in making the summer school a successful event.

Table of Contents

Multiple Access Communications Using Combinatorial Designs

Charles J. Colbourn

Computer Science, University of Vermont
Burlington, VT 05405
USA
Charles.Colbourn@uvm.edu

Abstract. In the past century, combinatorial designs have had substantial application in the statistical design of experiments and in the theory of error-correcting codes. Applications in experimental and theoretical computer science have emerged more recently, along with connections to the theory of cryptographic communication. This paper focuses on applications of designs in multiple access communications, in particular to balanced methods of sharing resources such as communications hardware, time, and bandwidth. The theory of combinatorial designs continues to grow, in part as a consequence of the variety of these applications and the increasing depth of the connections with challenging problems on designs.

1 Background

The nature of most computational tasks, including communication, is bursty. A single user often has an intensive requirement for a resource for a short period of time, followed by longer periods of relative inactivity. As a result, techniques for sharing resources have occupied many research areas within computer science and communications. Support for multiple access to a resource addresses both the needs of individual users for sufficient access to the resource, and the (economic) requirement that the resource itself be highly utilized. We focus here on problems of this type arising in multiple access communications; some of these applications have been described in the earlier surveys [37,39], and we borrow extensively here from those presentations.

Our interest is to explore the uses of combinatorial designs in the solution of such problems. The theory of combinatorial designs has a long and rich history. Its origins lie in somewhat specialized problems that arose in algebra, geometry, topology, and number theory. However, applications are found in the design of experiments [66] and in the theory of error-correcting codes [62]. Both fields of application served as sources for a wide variety of research directions.

In the past few decades, combinatorial design theory has grown to encompass a wider variety of investigations, many of which are not apparently motivated by any practical application. Rather they are motivated by a desire to obtain a coherent and powerful theory of existence and properties of designs. Nevertheless,

G.B. Khosrovshahi et al. (Eds.): Theoretical Aspects of Computer Science, LNCS 2292, pp. 1–29, 2002.
© Springer-Verlag Berlin Heidelberg 2002

it comes as no surprise that applications in experimental design and in coding theory continue to arise, and also that designs have found applications in new areas. Cryptography in particular has provided a new source of applications of designs, and simultaneously a source of new and challenging problems in design theory [72]. Across the spectrum of theoretical and experimental computer science, there are similar connections [37,39].

Arguably, many of the connections that arise are somewhat superficial, and appear to require only the translation of elementary combinatorial properties to the application domain. Naturally, the limited application does not then provide evidence of an important role for combinatorial design theory. However, we believe that there is ample evidence not only of superficial connections of theoretical investigations on designs to applications, but of deeper and more substantial connections. The importance of these cannot be overstated. While we can never know which results will find a genuine application, we expect to see such applications arise. Moreover, the evolution of the field depends largely upon its ability to make contributions both to other theories and to applications.

This paper presents some multiple access communication applications in which the connection with designs appears to be substantial. Our objective is to present evidence that combinatorial designs continue to arise in applications areas, often in unexpected ways; that the connections involve difficult aspects of the theory of designs; and that the applications motivate new research in design theory.

The applications span the breadth of computer communications. In this section, we introduce bus network design as a complement to the required definitions; then in Section 2 the design of multidrop or bus networks is examined in more depth. Multidrop or bus networks are used extensively in loosely coupled distributed systems; for tightly coupled systems, interconnection networks provide the basic communications structure. In Section 3, design-theoretic aspects of interconnection networks are briefly surveyed.

Section 4 then introduces a problem on graph designs and decompositions that arises in sharing wavelengths among point–to–point communication paths in an optical network such as SONET. In this application, communications are assigned to a specific portion of a wavelength so that no interference among communication arises. We then turn to systems that permit simultaneous transmission by multiple users, so that interference among users occurs. Section 5 considers how designs arise in schemes for multiple access that enforce a synchronous operation among users; Section 6 treats the analogous situation, but for asynchronous operation. Finally, in Section 7, superimposed codes for simultaneous transmission are described in the context of more general group testing problems.

These brief surveys of certain applications are independent of one another for the most part, and are intended to provide the reader with an appreciation for the breadth and depth of the relationships to combinatorial design theory.

We provide an overview of design theory next, but assume that the reader is familiar with the major topics in the area; see [19,36,40,46,59] for comprehensive treatments. In general, we attempt to give a sufficient introduction to the applications problem and then outline the connection with designs.

We provide enough background in combinatorial design theory to enable the reader to appreciate the role of combinatorial designs in the problems which we explore in subsequent sections. Many basic definitions are needed. We therefore introduce an illustrative recurring example to clarify the definitions as they are introduced. We italicize the paragraphs outlining the example, to distinguish them from the (authoritative) formal definitions.

Example. Imagine that there are v sites in a communications network. We are to establish communication paths among these sites. The communications links we have at our disposal are multipoint buses, or local area networks. Such a bus is attached to a number of sites, and is assumed to provide communication between any two sites attached to the bus. A bus network is a collection of buses, each attached to a subset of the network sites. The bus network is said to provide a direct connection between some subset of the network sites if there is a single bus in the network to which all sites in the subset are attached. For example, we could form a bus network with ten sites numbered 0 through 9, and six buses named A through F, with attachments as follows:

Bus	Sites on Bus
A	1,2,3
B	4,5,6
C	7,8,9
D	0,1,4,7
E	0,2,5,8
F	0,3,6,9

This bus network provides direct communication among nodes 0, 4 and 7 for example, since all three are attached to bus D. However, there is no direct communication here among sites 0, 4 and 5 despite the fact that direct communication can be established between any two of the three. Our first definition is a simple combinatorial model for such a bus network.

Let V be a finite set of v elements. A *set system* $\mathcal{B}$ on V is a collection of subsets of V; the set system is *simple* when $\mathcal{B}$ does not contain any subset more than once. Set systems are widely studied under the name *hypergraphs*. Sets in $\mathcal{B}$ are *blocks;* the number of blocks is denoted $b = |\mathcal{B}|$.

A bus network is just a set system in which V is the set of sites, and B contains a subset for each bus which contains precisely the sites attached to that bus. The bus network given earlier is a simple set system with v = 9 elements, and b = 6 blocks. In the design of actual bus networks, we are concerned with a number of criteria related to physical realizability and to performance. We list a few such criteria here.

1. *A site must not be attached to too many buses.*
2. *A bus must not be attached at too many sites (in order to maintain a manageable traffic load on the bus).*

3. *Every pair of sites must not appear together on too many buses (in order to avoid unnecessary redundancy).*

4. *Every pair, or more generally, subset of sites must appear together on sufficiently many buses (in order to ensure reliable direct communication).*

While (1) and (2) only specify upper bounds, if the buses are attached at too few sites, we need many buses, leading to a violation of criterion (1). Similarly, if a site appears on too few buses, we are forced to attach buses to many sites. This leads to a desire for "uniformity" or "balance" in determining the sizes of buses, number of buses attached to a site, and so on. Our next definitions therefore concern special types of set systems with these properties.

K denotes the set of *blocksizes* $\{|B| : B \in \mathcal{B}\}$. When K contains a single blocksize, k, the set system is *k-uniform*. Each element $x \in V$ appears in some subset $\mathcal{B}_x \subset \mathcal{B}$. The *replication number* $r_x > 0$ is $|\mathcal{B}_x|$, and the set of replication numbers is $R = \{r_x : x \in V\}$. Every subset $S \subseteq V$ appears as a (not necessarily proper) subset of a number of the sets in $\mathcal{B}$; the number of sets in $\mathcal{B}$ containing S is denoted $\lambda(S)$ and is termed the *index* of S in $\mathcal{B}$. The *t-index set* of a set system $(V, \mathcal{B})$ is the set $\{\lambda(S) : |S| = t, S \subseteq V\}$. A set system is *t-balanced* if the t-index set contains a unique value λ_t and $\lambda_t > 0$. A 1-balanced set system is simply one in which there is a single replication number r, with $r = \lambda_1$. Every set system is 0-balanced, with $b = \lambda_0$. Any t-balanced k-uniform set system is also $(t-1)$-balanced.

The blocksizes K are the bus sizes in our example. The replication numbers are the number of buses attached at each site, and the t-indices are the number of buses providing direct communication among subsets of t sites. Criteria (1)-(4) given earlier suggest that we choose a bus network which is both uniform and t-balanced. Let us consider an example of such a set system. We form a 4-uniform 3-balanced set system with $v = 8$ elements numbered 0 through 7, and 14 blocks. Each block is a 4-subset of elements; we adopt the usual convention of writing *ijkl* to denote the subset *i,j,k,l* whenever no confusion can arise. The block set $\mathcal{B}$ is 0124, 0137, 0156, 0235, 0267, 0346, 0457, 1236, 1257, 1345, 1467, 2347, 2456, 3567. So $k = 4$, $r = 7$, $\lambda_2 = 3$ and $\lambda_3 = 1$. What does this mean? It is a scheme for constructing a bus network on eight sites, having fourteen buses, with every bus attached to four sites, and every site on seven buses. Moreover, any pair of sites can establish direct communication via three different buses, and every set of 3 sites can communicate directly on a single bus! Set systems which exhibit this uniformity and balance address criteria (1)-(4) quite effectively. We focus on such special set systems.

A (balanced incomplete) *block design* is a pair $(V, \mathcal{B})$ where $\mathcal{B}$ is a k-uniform set system on V which is 2-balanced with index λ. $(V, \mathcal{B})$ is typically termed a (v, k, λ)-*design*. The remaining parameters, r and b, can be computed from the specified parameters, by observing the identities $vr = bk$ and $r(k-1) = \lambda(v-1)$. These identities are obtained by counting, in two ways each, the number of pairs that a given element appears in, and the total number of elements in the design, respectively.

Block designs are set systems in which the appearance of unordered pairs is uniform; the natural extension is to set systems which are t-balanced. A t-*design* $(V, \mathcal{B})$, or more precisely, a $t - (v, k, \lambda)$ design, is a k-uniform set system with $|V| = v$ which is t-balanced with t-index set $\{\lambda\}$. *Trivial* t-designs arise by taking $v = k$ or $k = t$.

Our last example is a 3-(8,4,1) design. Since it is also 2-balanced, the same set system is also an (8,4,3) block design. Given that $v = 8$, $k = 4$ and $\lambda = 3$, we compute $r = \lambda(v - 1)/(k - 1) = 7$ and $b = vr/k = 14$, as expected.

Now suppose in our example that there were nine sites rather than eight, and we therefore want a 3-(9,4,1) design to use as our bus network. Can such a design exist? Perform the following simple counting. Consider two sites, say x and y. Then for each other site z, the 3-subset x,y,z must be contained in precisely one block. There are $v - 2$ such 3-subsets containing x and y, and each block contains two such 3-subsets. But then if $v - 2$ is not even, there is no way to select blocks containing each of these 3-subsets once, and there is no 3-(9,4,1) design. We therefore need some necessary conditions to tell us for which parameters a design cannot exist, and also some conditions which guarantee existence.

Let us suppose that a $t - (v, k, \lambda)$ design $(V, \mathcal{B})$ exists. Consider a subset $X \subset V$ with $0 \le |X| < t$. The total number of t-subsets of V containing X in $\mathcal{B}$ is $\lambda\binom{v-|X|}{t-|X|}$, while any block containing X contains $\binom{k-|X|}{t-|X|}$ of these t-sets. Hence by considering all possible sizes of X, we obtain t divisibility conditions: $\binom{k-i}{t-i} \mid \lambda\binom{v-i}{t-i}$ for $i = 0, ..., t - 1$.

For example, for a $3 - (v, 4, 1)$ design to exist, we must have $4 \mid \binom{v}{3}$, $3 \mid \binom{v-1}{2}$ and $2 \mid v - 2$, from which we obtain the "congruence condition" $v \equiv 2, 4$ (modulo 6). Suppose a subset X is chosen, and let the subset of blocks in $\mathcal{B}$ each containing the set X be $\mathcal{B}_X$. $(V \setminus X, \mathcal{B}_X \setminus X)$ is a $(t - |X|)$-design, where $\mathcal{B}_X \setminus X$ is the set of blocks obtained by deleting from each block in $\mathcal{B}_X$ all points in X. This is the *derived* design for X.

A second type of necessary condition arises by considering the number of blocks. The number of blocks b in a block design is equal to $\lambda\binom{v}{2}/\binom{k}{2}$; a well-known inequality, Fisher's inequality, shows that $b \ge v$ in any block design, and hence $\lambda(v - 1) \ge k(k - 1)$.

Existence problems for block designs and t-designs are far from settled in general. We only summarize some main existence results here. For block designs, an elegant theory due to Wilson [80,81,82] establishes that the necessary conditions for the existence of a (v, k, λ) design are sufficient for v sufficiently large with respect to k. Hence existence of block designs is, in an asymptotic sense, well understood; nevertheless, complete solutions are known only for (v, k, λ) designs with $k = 3, 4$, and 5.

For t-designs with $t > 2$, much less is known. Teirlinck [76] proved that simple t-designs exist for all values of t. However, except for $3 - (v, 4, \lambda)$ designs, the necessary conditions are not known to be sufficient, even in an asymptotic sense. In fact, t-designs with index $\lambda = 1$ (*Steiner* systems) are at present unknown for $t > 5$. Much of the effort in combinatorial design theory has been invested in constructing designs with additional properties. Most effort to date in establishing existence results has been invested in triple systems $((v, 3, \lambda)$ designs), quadruple systems $(3 - (v, 4, \lambda)$ designs), and Steiner systems $(t - (v, k, 1)$ designs). We do not attempt to review this literature here; see [36].

In the bus network example, we have thus far ignored an important economic constraint. For practical reasons, we should add the constraint that the number of buses be as small as possible. We have already seen Fisher's inequality, which tells us that we need at least as many buses as sites, except in trivial cases. Can this be achieved? Before launching into more definitions, let us give a small example with seven points, 0 through 6, and seven blocks: 013, 124, 235, 346, 045, 156, 026. This is a (7,3,1) design, and is a 2-balanced bus network for seven sites using only seven buses. This is not an isolated example, as we see next.

A design with $b = v$ is a *symmetric* design. In a symmetric design, we have $\lambda(v - 1) = k(k - 1)$, and hence the parameters of a symmetric design are of the form $((k^2 - k)/\lambda + 1, k, \lambda)$; note that $b = v$ implies $k = r$. The *order* of a symmetric design is $n = k - \lambda$.

The case $\lambda = 1$ has received special attention. A symmetric design with parameters $(k^2 - k + 1, k, 1)$ is a *(finite) projective plane;* the parameters can equivalently be written as $(n^2 + n + 1, n + 1, 1)$, and the plane is then of order n (our example above is a projective plane of order 2). ¿From a symmetric design $(V, \mathcal{B})$, one can form a *residual* design by selecting one block, removing that block and removing all of its elements from the remaining blocks. Residuals of projective planes are $(n^2, n, 1)$ designs, usually called *affine planes*.

Projective planes can be obtained from a general class of structures which give rise to symmetric designs. Let S be an $(m + 1)$-dimensional vector space over $GF(q)$, the finite field with q elements, where q is a prime or prime power. The set of all subspaces of S is the *projective geometry of dimension m* over $GF(q)$, denoted $PG(m, q)$. The 1-dimensional and m-dimensional subspaces of S are *points* and *hyperplanes,* respectively. For each hyperplane H, let B_H be the set of points contained in H. Then using the 1-dimensional subspaces of S as points, the block set $\{B_H : H \subseteq S\}$ defines a symmetric design with parameters $v = \frac{q^{m+1}-1}{q-1}$, $k = \frac{q^m-1}{q-1}$, $\lambda = \frac{q^{m-1}-1}{q-1}$. Taking $m = 2$ yields projective planes. It follows that projective planes are known to exist for all values $n = q$ which are powers of primes; at present no planes of non-prime power order are known.

At the other extreme, the symmetric designs with parameters $(4n - 1, 2n - 1, n - 1)$ are *Hadamard designs,* and arise from related configurations, Hadamard matrices. Existence of Hadamard designs is still unsettled, but numerous infinite families of such designs are known. (The smallest open case is $n = 107$; see [36].)

We have already mentioned one reason for special interest in symmetric designs: they minimize the number of blocks in a design. A second reason is equally important. Two blocks in a symmetric (v, k, λ) design always intersect in precisely λ elements.

While we have seen many families of designs now which may be useful in designing bus networks, we have also seen that many parameter sets cannot be realized. Naturally, the impossibility of a certain set of parameters does not help us in solving the problem in designing bus networks. In such a situation, if we cannot have a design, we must relax one of the constraints on uniformity or balance. We could, for instance, allow buses of different sizes, but still require the t-balanced condition. On the other hand, we could insist that no t-subset appear too often, or that no t-subset appear too seldom. These generalizations lead to relaxations of block designs and t-designs.

In view of our observations that much remains to be settled concerning existence of designs, and that the necessary conditions rule out many orders, it is reasonable to try to "come close". Hence we might relax some of the restrictions. Suppose that $(V, \mathcal{B})$ is a set system which is k-uniform on v elements and each t-subset appears at most (at least) λ times in blocks of $\mathcal{B}$; then $(V, \mathcal{B})$ is a $t - (v, k, \lambda)$ packing ($t - (v, k, \lambda)$ covering, respectively).

A $t - (v, k, \lambda)$ packing can have at most $b(t, v, k, \lambda) = \lambda \binom{v}{t} / \binom{k}{t}$ blocks, while a $t - (v, k, \lambda)$ covering must have at least this number. Equality holds if and only if the packing (covering) is a $t - (v, k, \lambda)$ design. However, Rödl [68] demonstrates that as v goes to infinity, the size of a maximum $t - (v, k, \lambda)$ packing is $(1 - o(1))b(t, v, k, \lambda)$, and hence that we can always come "close" to a design.

Packings are also often called *partial* designs.

Packings and coverings relax the requirement that the index be constant; here we relax instead the requirement that the block size be constant. A *pairwise balanced design (PBD)* with parameters (v, K, λ) is a set system on v elements with blocksizes from K, and which is 2-balanced with index λ. A PBD does not in general have a unique replication number. Block designs are just PBDs with a single block size.

While there is a rich theory of pairwise balanced designs, we only remark on a few facts which we employ. First, Fisher's inequality applies to PBDs; hence symmetric designs are again PBDs with the minimum number of blocks. Second, Wilson's asymptotic existence theory applies to PBDs as well, and hence existence of a desired PBD is assured for v sufficiently large, provided that basic numerical conditions are met. Finally, one can both relax restrictions on blocksizes and impose only an upper bound on the index; the result is a partial PBD (equivalently, a non-uniform packing).

To illustrate these last definitions, consider a bus network connecting ten sites numbered 0 through 9. Let us suppose that no bus is attached at more than four sites, and we require a unique direct connection between every pair of sites. Employing a (10,{3,4},1) PBD with blocks 0123, 0456, 0789, 147, 258, 369, 159, 267, 348, 168, 249, 357 gives a design for such a bus network. Alternatively, if each bus is attached at most three sites, replacing blocks 0123, 0456, 0789 by blocks 012, 123, 045, 456, 078, 789, 036, 039 gives a (10,3,1) 2-covering. This can be used to establish at least one direct connection between every pair of sites.

In our running example, we have found motivation for most of the basic concepts in design theory. Two further basic properties are needed in other applications, but do not arise in the bus network scenario. We introduce these next.

One property of designs which arises in numerous design applications deserves special attention. For a design $(V, \mathcal{B})$, a *parallel class* (or *resolution class*) of blocks $\mathcal{P} \subset \mathcal{B}$ is a set of blocks such that no two intersect, and the union of all blocks of $\mathcal{P}$ is V; a *near-parallel class* is similar, but the union contains all but one element of V.

When $\mathcal{B}$ can be partitioned into parallel classes, this partitioning is a *resolution,* and $(V, \mathcal{B})$ is a *resolvable* design.

More generally, a $t - (v, k, \lambda)$ design may be *partitionable* into $(t') - (v, k, \lambda')$ designs; resolution is just the case $t' = 1$ and $\lambda' = 1$. Applications arise for partitionable designs, but especially for the restricted case, resolvable designs.

Often we choose a design with some symmetry, or nontrivial automorphism. An *automorphism* of a design is a bijection from the elements onto themselves, by which blocks are mapped to blocks, and subsets which are not blocks are mapped to subsets which are not blocks. Provided that an automorphism is known, the design has a compact representation, in which a representative for each equivalence class (orbit) of blocks under the action of the automorphism is retained. The existence of such a compact representation enables one to find and use much larger designs in practical applications.

Of particular interest are designs with a *cyclic* automorphism, which is a cycle involving all elements of the design. Colbourn and Mathon [42] provide a survey on cyclic designs, and remark on the importance of the compact representation here; an instructive example of the use of cyclic designs appears in Chung, Salehi and Wei [31] and in Section 6.

2   Multidrop Networks

Our running example in the introduction, the design of bus networks, forms one of the clearest illustrations of how combinatorial designs arise in the allocation of resources. Indeed, beyond the simple connections already mentioned, there are deeper ones. Hence in this section, the design of networks using broadcast media so that every two sites lie on a common link, subject to constraints on the number of links at each site (degree), and the number of sites on each link (link size), is examined. This leads to the examination of pairwise balanced designs and to the use of (k, n)-arcs in projective planes.

The network design problem of interest is as follows. There are n network sites, to be connected using multidrop communication links such as Ethernets, token rings, or any broadcast medium. A *link* or *bus* is a subset of the n sites. In order to avoid congestion due to switching overhead from one link to another, it is required that every two sites appear together on at least one link. Typically, each site is equipped with a limited number of communication ports and hence can appear on at most some fixed number r of the links. Similarly, each link has a limit on the number of sites that it can connect. Reasons for such a limitation include capacity limits, and limits on acceptable routing delay within the link. With these constraints in mind, the problem can be informally stated as follows: Connect n sites so that every two sites appear together on at least one link, subject to the constraint that no link has more than k sites on it, and no site appears on more than r links.

Problems of this type have been studied extensively. Mickunas [65] considered the case when k and r are close to equal. Subsequently, Bermond and his colleagues [14,15,16] considered general network design problems of this type under the name "bus interconnection networks". They are primarily responsible for observing that numerous well studied combinatorial configurations lead to useful solutions to such network design problems; see also [5,39,44,85,90,91].

Block designs and pairwise balanced designs lead to optimal solutions for the network design problem when $k < r$ [14,15]. When $k = r$, projective planes yield bus networks [65]. Practical concerns dictate that the replication number r be a fixed small number, while the block size k can be potentially much larger than r. Since block designs and PBDs always have $k \leq r$ by Fisher's inequality [19], a technique is needed to treat cases when $k > r$. This is one of the problems treated in [86].

Bermond, Bond, Paoli, and Peyrat [14] propose the following. Suppose that we are to construct a covering with replication number at most r and block size at most k, and our objective is to maximize the number of elements. Choose q so that q is a power of a prime, $q + 1 \leq r$, and q is as large as possible subject to these constraints. Then form $\text{PG}(2,q)$ on element set V of size $q^2 + q + 1$, and with block set $\mathcal{B}$. A *weight function* $\omega : V \to \mathbb{Z}^+$ from elements to positive integers is to be chosen, and a set of elements $W = \{(x, i) : x \in V \text{ and } 1 \leq i \leq \omega(x)\}$ defined. The weight of an element indicates the number of times that it is replicated in W. We choose ω so that the *weight*, $\omega(B)$, of block B satisfies $\omega(B) = \sum_{x \in B} \omega(x) \leq k$, for every $B \in \mathcal{B}$, and defines a new set of blocks

$$\mathcal{D} = \{\{(x, i) : x \in B \text{ and } 1 \leq i \leq \omega(x)\} : B \in \mathcal{B}\}.$$

Then $(W, \mathcal{D})$ is a covering with block sizes at most k, replication number $q+1 \leq r$, and $\sum_{x \in V} \omega(x)$ elements. Naturally the problem is to determine ω so as to constrain the weight of each block to k while maximizing the number of elements. Bermond et al. [14,15] conjectured that the covering with replication number at most r, block sizes at most k, and the largest number of elements, arises in this manner when $r - 1$ is a prime power. It follows from a theorem of Füredi [55] that such a covering can have at most $rk - (r-1)\lceil \frac{k}{r} \rceil$ elements. Now choosing ω so that all element weights are as equal as possible subject to the constraint on block weight leads to coverings for which Füredi's bound is achieved infinitely often, and approaches this bound as $k \to \infty$ for fixed r when $r - 1$ is a prime power [15,16]. Hence, although there are many potential methods for producing coverings, Füredi's result establishes that the asymptotically optimal coverings arise from replicating elements in projective planes.

Bermond et al. [15,16] did not address the question of finding the largest number of elements in a covering with block size at most k and replication number at most r precisely. Yener, Ofek, and Yung [86], however, employed a similar underlying strategy but developed techniques for specifying the weight of each element in the projective plane so as to maximize the number of elements. Their method is a simple greedy strategy, which does not in general lead to the minimum block size [35]. See [20] for related work on affine planes.

When $k \leq r$, the existence of bus networks with restricted bus sizes amounts simply to the existence of pairwise balanced designs with specified block sizes. When $k > r$, two more specific problems arise.

The first arises when r is of the form $q + 1$ for q a prime power, so that a projective plane of order q exists. Then an apparently hard question is to determine the minimum increase in the maximum block size. A (k, n)-*arc* in a projective plane of order q is a nonempty set K of k elements such that n is the maximum number of elements in K that appear together on a block. A $(k, 2)$-arc is a k-*arc*. The existence of (k, n)-arcs was extensively studied, but their importance here is that the maximum number $m_n(q)$ of elements in a (k, n)-arc in $PG(2, q)$ is precisely the same as the maximum number of elements that can be replicated without increasing the maximum block size by more than n.

Barlotti [9] established that $m_n(q) \leq (n-1)(q+1)+1$. A simple computation establishes that when $PG(2, q)$ contains a (k, n)-arc whose size meets this bound, replicating the elements of the arc yields equality in the asymptotic bound of Füredi [55] discussed earlier. Unfortunately, the determination of $m_n(q)$ is a very difficult problem in finite geometry that remains far from settled [57,58]. Except when $n = q + 1$, equality in Barlotti's bound can be achieved only when n is a divisor of q [9], and is achieved in two *trivial cases*: when $n = 1$ (by a single element), and when $n = q$ by all elements not lying on a fixed block. When q is a second or higher power of a prime, nontrivial arcs meeting Barlotti's bound *always* exist when q is a power of 2 and n is a divisor of q [45]. However, in a 1997 breakthrough it has been shown that they *never* exist when q is a power of an odd prime [8].

When $m_n(q)$ does not realize Barlotti's bound, extensive research has attempted to obtain lower and upper bounds, and specific exact values; as an introduction to the literature, we suggest [57,58]. Each lower bound can lead to a replication scheme for producing a covering, and each upper bound establishes a limit on how well such a replication scheme can do.

The second main problem arises when r is not one more than a prime power. Then the advisability of beginning with a plane of order less than r is open to question. Colbourn [35] observed that certain related combinatorial configurations can lead to better results. Consider, for example, the case when $r = 7$ and $v = 39$. There is no plane of order 6, and hence the plane $PG(2, 5)$ can be used. Then block size 9 is obtained. However, there is a covering on 39 elements with replication number 7 and block size 7 [79], and hence in this case replication of elements in planes does not appear to lead to the best solution. For this reason, we mention a less well studied generalization of difference sets that can lead to (slightly) smaller block sizes for certain degree constraints.

A *difference cover* modulo v of *order* q, $D = \{d_0, \ldots, d_q\}$, has the property that $\{d_i - d_j : 0 \leq i, j \leq q \text{ and } i \neq j\}$, arithmetic modulo v, contains every nonzero integer in $\mathbb{Z}_v$. Hence every nonzero difference arises at least once as the difference modulo v of two elements in D. When $v = q^2 + q + 1$, a difference cover is a difference set. However, while difference sets only exist for certain values of q, difference covers exist for every value of q. Of course, the price is that the number of elements v is less than $q^2 + q + 1$ in general. Now adding each integer i to the elements of D in turn, we produce v blocks forming a covering with block size $q + 1$ and replication number $q + 1$. Wiedemann [79] gave a difference cover modulo 39 of order 6, which provides the illustration given above. He also presented a table of the smallest order difference cover modulo v for each value of $v \leq 133$. Unfortunately, these computational results are not at present accompanied by a useful theory. For certain small replication numbers such as 7 (a difference cover modulo 39) and 11 (a difference cover modulo 95 [79]), it appears that difference covers can improve upon the use of planes. Difference covers arise in an apparently unrelated application, the design of quorums for ensuring mutual exclusion [38].

We have focussed on the design of bus networks with diameter one. However, transversal designs arise also in the design of bus networks of larger diameter; see [13], for example.

3   Channel Graphs and Interconnection Networks

This section examines interconnection strategies for computer networks and the input-output channel graphs that they contain.

The generic *interconnection problem* is to establish communication paths between *input nodes* V_I and *output nodes* V_O; we allow the case that the input nodes and output nodes are the same. *Switch nodes* V_S may be used; these simply relay a message from one communications channel to another. Our task is to connect the nodes in $V = V_I \cup V_O \cup V_S$ using point-to-point *links*. The network must be *connecting*, in that there must be a communication path from each input node to each output node. The *distance* from an input node to an output node is the number of links in the shortest communication path connecting them. The *diameter* or *depth* is the maximum distance from an input to an output. To ensure small delay in communication, we require that the diameter be small.

When all input-output paths have length equal to the diameter, the switch nodes (if any) can be partitioned into *stages*, by placing all nodes at distance i from an input in the ith stage. Networks admitting such a partitioning are *multistage interconnection networks*. A multistage interconnection network with a single input node and a single output node is a *channel graph*. In the analysis of interconnection networks, the network designer is concerned with the overall network design. However, each input-output pair is concerned with the portion of the network containing the paths from the specified input to the specified output, i.e. the input-output *channel graph*.

In some applications, we require that the n input nodes can simultaneously communicate with the n output nodes, given a specified mapping of inputs to outputs. This requires *disjoint* communication paths, which share no common link or intermediate node. A good example of this situation arises in the design of shifting networks.

A *barrel shifter* is a network whose nodes are $\{0, 1, \dots, n-1\}$, the integers modulo n. Given a shift distance s, $1 \leq s < n$, every node must transfer a value to the node whose label is s larger; more precisely, for each $0 \le i < n$, node i must establish a connection to node $i + s$ (reduced modulo n as needed), and all n communication paths are to be disjoint. Kilian, Kipnis and Leiserson [61] developed a barrel shifter which has diameter one; when implemented in VLSI, the shift is accomplished in a single clock cycle.

We now consider an even stronger connection property of networks. An n-superconcentrator is a network with n inputs and n outputs in which disjoint communication paths can be established from the inputs to the outputs in *any* of the $n!$ possible orderings. We restrict superconcentrators to have only links, and no larger buses. A superconcentrator of depth one requires all n^2 connections (i.e. each input connected to each output); hence superconcentrators of depth greater than one are of interest. Nevertheless, superconcentrators are typically constructed using special types of depth-one networks in which every set of inputs is directly connected to a relatively large set of outputs (see, for example, [29]). More formally, a network $(V_I \cup V_O, E)$ with $V_I \cap V_O = \emptyset$ is a (n, α, β)-*expander* if every set of α inputs is directly connected to at least β output nodes.

The motivation for strong expansion capability is to avoid congestion or blocking. To quantify the disruption due to such blocking, examine the channel graphs for each input-output pair. We assume that the probability q_i of occupancy of each edge in a channel graph is known. The *blocking probability* of a channel graph is defined as the probability that every channel of that graph contains at least one occupied (blocked) edge. A channel graph with k stages is *superior* to another channel graph with k stages if the blocking probability of the former never exceeds that of the latter, independent of the occupancies for the E_i. There is no guarantee that constructing an interconnection network in which channel graphs are superior leads to an interconnection network with high expected throughput. Nevertheless, if the channel graphs have high blocking probability, this ensures poor throughput. Hence the design of superior channel graphs arises as a necessary step in interconnection network design.

A network of diameter one is a 2-$(n, K, 1)$ covering; if we require in addition that each node i has a (disjoint) path to node $i + s \bmod n$, the n pairs from the set $D_s = \{\{i, (i + s) \bmod n\} : 0 \le i < n\}$ must appear in n distinct blocks. At first, this seems to be a complicated requirement, but a widely studied class of designs always has the desired property; we introduce them here. A set system $(V, \mathcal{B})$ with $V = \{0, 1, \dots, n - 1\}$ is *cyclic* if, whenever $\{b_1, \dots, b_k\} \in \mathcal{B}$, $\{b_1 + 1, \dots, b_k + 1\} \in \mathcal{B}$ (arithmetic modulo n is used). The *orbit* $\mathcal{O}(B)$ of a block B is the set $\{B + s \pmod{n} : 0 \le s < n\}$; it is *full* when $|\mathcal{O}(B)| = n$. When all orbits are full, the set system is *full-cyclic*. It is easy to see that the pairs of D_s appear in at least n distinct blocks of a full-cyclic covering.

Any full-cyclic covering can then be used to design a barrel shifter. Each node finds the first orbit in which $\{0, s\}$ appears, say in block B. Node x now writes its value to the bus $B + x \pmod{n}$, and reads its value from the bus $B + x - s \pmod{n}$. In this way each node x reads the value node $x - s \pmod{n}$ wrote, and each communication path corresponds to a unique block in the orbit. Kilian, Kipnis and Leiserson [61] observed that to minimize the total number of buses and the number of buses incident at a node, the covering chosen is a *cyclic projective plane* (i.e. a projective plane which is cyclic).

The actual operation of a barrel shifter based on a cyclic projective plane is remarkably simple. To see this, consider the structure of cyclic projective planes. Since there are only n blocks, any two blocks B_1, B_2 satisfy $B_1 \equiv B_2 + s \pmod{n}$ for some $0 \leq s < n$. Consider a single block $B = \{b_1, \ldots, b_k\}$. Now for each element d, $1 \leq d < n$, $\{0, d\}$ appears in exactly one block. Hence B must contain exactly two elements b_i, b_j for which $b_j - b_i \equiv d \pmod{n}$. Every d, $1 \leq d < n$, is the difference of two elements of B; such a set B is a *difference set* for $\{0, 1, \ldots, n-1\}$.

Using the difference set representation of the cyclic projective plane, the operation of a barrel shifter is straightforward. To shift a distance of s, each node finds the two elements b_i, b_j in the difference set with $b_j - b_i \equiv s \pmod{n}$. Node x then writes onto bus $x + b_i$, and reads from bus $x + b_j$.

When no cyclic projective plane on n elements exists, this very simple control logic can be retained nonetheless: This scheme requires only a set which covers all differences from 1 to $n-1$. Hence we can use a *difference cover*, in which each d, $1 \leq d < n$, is the difference of at least one pair of elements. Kilian, Kipnis and Leiserson [61] observed that they produce optimal barrel shifters of depth one. They also use difference covers to design 'permutation architectures', which realize permutations other than just cyclic shifts.

Let us turn to superconcentrators. Any depth-one network with $V_I \cap V_O = \emptyset$ can be equivalently written as a set system $(V_I, \mathcal{B})$, where $\mathcal{B} = \{\{v_i : \{v_i, v_o\} \in E\} : v_o \in V_O\}$. In this setting, an (n, α, β)-expander is a set system with n elements and n blocks, so that every set of α elements intersects at least β of the blocks. Intuitively, β is largest when the blocks intersect each other as little as possible. At the same time, however, for β to be large, each element must appear in a large number of blocks. To maximize the expansion, we choose to balance the block sizes, and balance the sizes of block intersections. Hence we consider symmetric designs.

Alon [1] proves that one class of symmetric designs, obtained from the points and hyperplanes of the projective geometries $PG(d, q)$, provides good expansion properties: In the design from $PG(d, q)$ on n elements, every set of α elements intersects $\beta \geq (\alpha n)/(\alpha + q - 1)$ blocks. Hence for all $\alpha = o(n)$, $\alpha = o(\beta)$; such a network is termed *highly expanding*. Moreover, Alon remarked that these expanders have essentially the smallest number of links of any network with equivalent expansion properties. Using projective geometries for expanders, Alon [1] established the existence of n-superconcentrators of depth three with $O(n^{4/3})$ links; see [1] for further uses of the expanders and superconcentrators, and [56] for a similar use of symmetric designs.

In the same way that designs lead to desirable expansion properties, they also arise in the design of superior channel graphs. Chung [28,30] used incidence graphs of block designs to determine connections between the second and third stage of a four-stage channel graph. She established that, among all channel graphs with the same numbers of nodes in each stage and the same numbers of interstage links, channel graphs arising from block designs are superior.

The design of interconnection networks employs design-theoretic tools in a number of ways. The use of designs to cover all pairs of nodes is prevalent in diameter one networks; on the other hand, the balanced intersection of blocks is shown to lead to high expansion factors, and hence to highly connecting networks.

4   Ring Grooming in SONET Networks

In this section, we outline an application of combinatorial designs to wavelength assignment in SONET networks [27,71]. A *synchronous optical network (SONET) ring* on n sites is an optical interconnection device. The sites are arranged circularly. A *clockwise* or *right ring* connects the ith site to the $(i + 1)$st, and a *counterclockwise* or *left ring* connects the ith site to the $(i - 1)$st. This provides two *directions* in which traffic can be delivered between any two sites.

Each optical connection can carry multiple signals on different wavelengths. However, the number of wavelengths is limited, and the bandwidth on each wavelength is also limited. Typically, one goal is to minimize the number of wavelengths used. An equally important goal is to ensure that each wavelength has sufficient bandwidth for the traffic it is to carry.

Now let us examine a combinatorial problem. Let n be a positive integer. Let $\mathbb{Z}_n$ denote the set $\{0, \ldots, n-1\}$; when arithmetic is done on elements of $\mathbb{Z}_n$, it is carried out modulo n. Let $\mathcal{P}_n$ be the set $\{(i,j) : i,j \in \mathbb{Z}_n, \ i \neq j\}$. Partition the sets of $\mathcal{P}_n$ into two sets, $\mathcal{L}$ and $\mathcal{R}$. Then associate with each pair $P = (i,j) \in \mathcal{P}_n$ the set $S(P) = \{i, i+1, \ldots, j-1\}$ if $P \in \mathcal{R}$, or the set $S(P) = \{j+1, \ldots, i-1, i\}$ if $P \in \mathcal{L}$.

Partition $\mathcal{P}_n$ into s classes, $C_1, \ldots, C_s$. Compute the multiset union $M_i = \bigcup_{P \in \mathcal{R} \cap C_i} S(P)$ and the multiset union $N_i = \bigcup_{P \in \mathcal{L} \cap C_i} S(P)$. Let g be an integer. If, for every $1 \leq i \leq s$, the multisets M_i and N_i do not contain any symbol of $\mathbb{Z}_n$ more than g times, then $C_1, \ldots, C_s$ is an (s, g)-*assignment* for the partition $\mathcal{L}, \mathcal{R}$ of $\mathcal{P}_n$.

Since we are at liberty to choose the partition of $\mathcal{P}_n$ into $\mathcal{L}$ and $\mathcal{R}$, we define an (n, s, g)-*assignment* to be the partition together with the (s, g)-assignment for that partition. We shall be concerned with those (n, s, g)-assignments that minimize s for particular values of n and g. Among these assignments, we prefer certain ones realizing a minimality condition, described next.

Consider a particular (n, s, g)-assignment. Let $V_i = \{x, y : (x, y) \in C_i\}$. The *drop cost* of the assignment is defined to be $\sum_{i=1}^s |V_i|$. For specific choices of n and g, what is the smallest value of s that we can achieve? For (n, s, g)-assignments, what is the smallest drop cost that we can achieve? We address these two questions, and describe an approach that uses techniques from combinatorial design theory and graph decompositions to obtain results on the existence of such assignments.

Let us examine how the basic SONET application is modeled in this combinatorial formulation. The sites of the SONET ring are the elements in $\mathbb{Z}_n$. Then $\mathcal{P}_n$ simply indicates the pairs of sources and destinations for communication. The choice of left or right ring on which traffic is to be routed is indicated by the partition into $\mathcal{L}$ and $\mathcal{R}$. Then for $P \in \mathcal{P}_n$, the set $S(P)$ is precisely the originating site together with all of the intermediate sites through which the traffic flows for the pair P. The number of wavelengths used is s, and the partition into classes $C_1, \ldots, C_s$ specifies the chosen wavelengths.

Suppose that the pairs $P_1, \ldots, P_t$ have all been assigned to the same direction and the same wavelength. Suppose further that some site $h \in \mathbb{Z}_n$ has the property that $h \in \bigcap_{i=1}^{t} S(P_i)$. Then all traffic involving these source–destination pairs must be routed through site h. If the traffic requirement for these t pairs exceeds the bandwidth, then site h would be unable to handle the traffic. In the absence of specific information about the traffic requirements, we suppose that no pair has a traffic requirement exceeding $\frac{1}{g}$ times the bandwidth. Then the condition on M_i and N_i ensures that sufficient bandwidth is available on each wavelength in each direction.

In a communication, the source and destination sites typically convert between the electrical and optical domains, while intermediate sites are all-optical forwarding devices. To start or terminate a connection is more expensive. *Wavelength add-drop multiplex* (*WADM*) permits a wavelength to bypass a node without the costly termination when no traffic on the wavelength originates or terminates at the node. Hence costs of a SONET ring configuration using WADM can often be lowered by reducing the number of different source and destination sites on each wavelength. The drop cost of the assignment defined earlier gives the number of SONET ADMs employed, and our interest is to minimize this number.

We have outlined an optical communications environment in which traffic from a particular source to a particular destination remains on a single wavelength. For arbitrary (s, g)-assignments, we cannot always partition each wavelength into g channels so that that each source–destination pair remains on a single channel. Nevertheless, the implementation is substantially simplified when we can do so. If an (s, g)-assignment can be partitioned into channels in this way, then it arises from an $(sg, 1)$-assignment by forming s unions of g classes each. This special type of assignment is a *grooming*. We shall typically require that the assignments produced are, in fact, groomings.

The case when $g = 1$ arises when each communication requires the entire bandwidth available on a wavelength. Two pairs P and P' can nonetheless share the bandwidth if they are on opposite (left and right) rings, or if they are on the same ring and $S(P) \cap S(P') = \emptyset$. Our task is then to partition the set $\mathcal{P}_n$ into s wavelengths and two directions, so that within each we find each site at most once. It is then necessary that any such partition of the pairs $\mathcal{P}_n$ also partition the multiset $\mathcal{M} = \bigcup_{P \in \mathcal{P}_n} S(P)$ into s classes and two directions, each containing every symbol in $\mathbb{Z}_n$ at most once. Now $S(P)$ depends upon the direction chosen for P, but let us suppose for the moment that $S(P)$ contains at most $n/2$ elements. This can be guaranteed *if* we choose the shorter direction around the ring. Then $\mathcal{M}$, by an easy counting argument, has $(n^3-n)/4$ elements when n is odd, and $n^3/4$ when n is even. Since each wavelength accounts for at most $2n$ of the entries in $\mathcal{M}$, we require that $s \geq \lceil \frac{n^2-1}{8} \rceil$ wavelengths be available when n is odd, and $s \geq \lceil \frac{n^2}{8} \rceil$ when n is even. See, for example, [17,71].

Wan [78] describes a very useful set of "primitive rings" which provide the wavelength assignment. We review his method briefly. First suppose that $n = 2m$ is even. Define the rings $Q_{ij}^n = \{(i,j), (j, i+m), (i+m, j+m), (j+m, i)\}$; when these rings are routed clockwise and $i < j < i+m$, the pairs of each ring can be placed on a single wavelength. Next define the rings $R_i = \{(i, i+m), (i+m, i)\}$.

We place the pairs in the rings Q_{ij}^n in $\mathcal{R}$ when $0 \leq i < j < m$. Then, for $0 \leq i < \lceil n/4 \rceil$, we place the ring R_i in $\mathcal{R}$. All other pairs are placed in $\mathcal{L}$. Indeed, when Q_{ij}^n is placed in $\mathcal{R}$, we suppose that Q_{ji}^n is placed in $\mathcal{L}$. The rings Q_{ij}^n and Q_{ji}^n involve the same sites and can be placed on the same wavelengths in opposite directions.

It is easily verified that such a set of primitive rings minimizes the number of wavelengths (indeed, every wavelength is used at every single site). When $n = 2m+1$ is odd, a simple variant can be used. Begin with the primitive rings described for the even order $n-1$. We introduce a new symbol ∞ between $n-1$ and 0. Rings Q_{ij}^{n-1} are placed as before. However, the rings R_i are removed, and each is replaced by two rings, $T_i^n = \{(\infty, i), (i, i+m), (i+m, \infty)\}$ in $\mathcal{R}$ and $T_{i+m}^n = \{(\infty, i+m), (i+m, i), (i, \infty)\}$ in $\mathcal{L}$.

Using this solution for $g = 1$, one way to produce an assignment with $g > 1$ is to combine up to g primitive rings, while minimizing number of wavelengths and drop costs. A method for forming an assignment by taking unions of primitive rings is a *grooming* of primitive rings. The most natural case to treat is when g is a power of two as a result of the bandwidth hierarchy available in SONET rings.

Wan [78] solves the cases when $g = 2$ and $g = 4$, and develops some general techniques. We review his method briefly. First, let $n = 2m+1$ be odd. Form a graph LK_m, a complete graph with a loop on each vertex. Associate with edge $\{i, j\}$, $i < j$, the primitive ring Q_{ij}^{n-1}, and with each loop $\{i\}$ the primitive ring T_i^n. Now choose a subgraph of LK_m having q edges and p vertices of nonzero degree. If we place the primitive rings corresponding to these edges in the clockwise direction on the same wavelength, the bandwidth suffices *exactly when* $g \geq q$. Indeed, this simply restates the requirement that every site be involved as a sender or intermediate site in at most g primitive rings on the same wavelength and same direction. What about the drop cost? Each edge $\{i, j\}$ with $i < j$ corresponds to a primitive ring involving four sites, while each loop corresponds to a primitive ring involving three. When a vertex appears in more than one edge, a savings results. When no loops are chosen, the *cost* for this wavelength (or subgraph) is easily calculated to be $2p$; when at least one loop is chosen, it is $2p+1$.

When n is even, the situation is fundamentally the same, but some details differ. We examine this next. While for the case of odd n, the primitive rings fall naturally into pairs, the primitive rings R_i do not 'naturally' pair when n is even. Indeed, if we select a graph in the decomposition whose number of nonloop edges plus *half* the number of loops does not exceed eight, then we can make a valid assignment to wavelengths as follows. On the clockwise ring, we place a primitive ring of size four, placing its reversal on the counterclockwise ring. However, for primitive rings of size two, we place half on the clockwise ring and half on the counterclockwise ring. Hence in the even case we treat loops differently than when the ring size is odd. In this case, the *cost* of a subgraph is always simply twice the number of vertices of nonzero degree in the subgraph.

Our task in both cases has been reduced to a *graph decomposition problem*. Partition the edges of LK_m into subgraphs, each containing at most g edges (counting loops as edges when the ring size is odd, and as half-edges when the ring size is even) so that the number of subgraphs is minimized *and* so that the total cost of all chosen subgraphs is minimized. See [36,67] for results on graph decompositions in general and for further references.

5   Synchronous Multiple Access to Channels

The next section examines an application involving sharing an optical channel. In that context, users act asynchronously, and hence the channel decoding involves cyclic shifts of codewords. This section examines an analogous problem. In this variant, however, rather than multiplexing by partitioning the channel's capacity into discrete time slots, we employ multiplexing based on available frequencies. This avoids some of the issues that arise in synchronization, but introduces some additional complexities.

Each user is to be able to send one of m different messages in a channel, or can remain silent. The channel is capable of carrying any subset of v different pulses or tones simultaneously, and can be equipped with intensity detection devices that determine not only the presence of a particular tone, but also the intensity with which this tone was employed. The latter is usually measured in multiples of some basic nonzero intensity, and accurately distinguishes large variations in intensity. However, small variations are not considered to be significant, in order to allow for noise.

Each message for each user is mapped to a *codeword*, which indicates a selection of k of the v available tones (i.e. the scheme is *multi-tone*). When the transmitter is silent, no signal is sent. When active, the transmitter sends the combination of k tones corresponding to the desired message. Typically, few transmitters are active. As with optical orthogonal codes, a receiver must be able to detect the presence of a message from a particular transmitter. For this reason, interference resulting from the simultaneous transmission of two (or more) codewords is to be kept to a minimum. When, for example, every transmitter is assigned only one message, we require that codewords for two different users share at most one tone. If all users are assigned one message only, and there is no intensity detection, then the task of the receiver is precisely that of solving a nonadaptive group testing problem (see Section 7). Intensity detection enables us to determine (with some degree of accuracy) the number of users who have transmitted a particular tone. Hence the problem is the variant of nonadaptive group testing in which tests report not just the presence of a defective, but also the number of defectives (see [49, Chapter 5]). Codes for this type of "spread spectrum" signalling system are described in [3,4,52,53].

The m-ary problem is described in some detail in [88,89]. Transmitters have a collection of m different messages and can choose any one to send or remain silent. The intended application here is to signalling systems in which, despite the large number of users, traffic from each user is bursty (i.e. high volume but short duration). The design of the system optimizes the handling of traffic when a single user is active, but permits multiple access by a small number of users. We retain the requirement that codewords assigned to different users share at most one tone. We enforce in addition a requirement that two codewords associated with different messages used by the same transmitter share no tone at all. This "orthogonality" requirement permits the most accurate decoding when a single user is active.

To accommodate multiple users, again it is necessary to be able to determine which combination of messages is present in the channel. For this reason, it is generally considered to be a poorer signalling design if one tone is used much more often than another. A secondary, but still important, criterion is therefore that all available tones appear in approximately the same number of assigned codewords.

When every transmitter has a unique codeword, a simple design theoretic problem arises. Associate with each of the v tones an element, and with each codeword a block which is a subset of k elements. Blocks then have the property that they intersect in at most one element, and hence no pair occurs in more than one block. The result is a packing of index one, block size k, and order v. Maximizing the number of transmitters requires simply the choice of a maximum packing.

However, even this basic signalling problem poses some difficulties. When multiple transmitters are active concurrently, the received signal is the union of the transmitted signals. With intensity detection, the received signal is the multiset union. In the former situation, our task is to recover from the union the constituent sets; as noted, this is a nonadaptive group testing problem. It is not known whether, by assuming the availability of multiset unions rather than of set unions, better packings can be found. Codes can be found for which multiset unions permit proper reception while set unions do not; however, in the case $k = 3$, the largest number of codewords can be realized by a code for which set unions suffice [34].

The multi-tone systems lead to difficult problems in design theory as well. We concentrate on the case when blocks are triples. A code to be used in a signalling system is necessarily a packing by triples. Suppose that the packing to be used is on v elements and has b triples. If every user is to be assigned m of the triples, we require that the m triples assigned form a partial parallel class (i.e. any two triples in the class are disjoint). Then $m \leq \lfloor v/3 \rfloor$, and the maximum number of users that can be supported cannot exceed $\lfloor b/m \rfloor$. A suitable code for s users consists of a packing on v elements with ms triples, partitioned into s partial parallel classes of size m. Among such packings, those in which every two elements appear in approximately the same number of triples are preferred.

In the case $m = \lfloor v/3 \rfloor$, these packings have been extensively studied. For example, when $v \equiv 3 \pmod 6$, the solutions are Kirkman triple systems (i.e. resolvable Steiner triple systems) [89]. Indeed, the "frame" obtained by deleting a single element in a Kirkman triple system provides a solution when $v \equiv 2 \pmod 6$. When $v \equiv 0 \pmod 6$, nearly Kirkman triple systems provide solutions, and when $v \equiv 1 \pmod 6$, Hanani triple systems provide the codes; see [40] for more details about these types of triple systems.

Colbourn and Zhao [41] completed the solution when $v \equiv 4, 5 \pmod 6$, so that the determination of codes when the number of messages is maximum is complete. In the intended application, although it is plausible that the number of messages coincides with the maximum permitted, this is unlikely. The primary application is in systems employing a digital to analog conversion, so that a chunk of ℓ bits in an incoming datastream is converted to a message in the form of the k tones selected. Typically, then, we find that $m = 2^\ell$, so that m is a power of two. Zhao [88] observed that, beginning with a packing partitioned into maximum partial parallel classes, simple heuristics usually suffice to partition the same packing into more and shorter partial parallel classes. Despite this evidence that partitioning is more difficult when m is large, and the use of such partitions in forming codes for smaller m, the current state of affairs in our knowledge of triple systems is quite incomplete. To begin with, the existence of solutions for large values of m does not always ensure the existence of solutions for smaller values of m. The Kirkman triple system of order 9, for instance, admits a partition with 4 classes of size 3, but does not admit a partition with 6 classes of size 2. More importantly, no method with a performance guarantee appears to be available at present which permits us to massage a packing with large partial parallel classes into one that has smaller but more partial parallel classes. Indeed, the existence question given m, s, and v asking for a packing by triples on v elements and ms blocks which has a partition into s partial parallel classes of v blocks has been solved only for certain restricted cases [41,88,89]. Among these are included the cases for all small values of v when m is a power of two, which form the principal cases employed in the application.

6 Optical Orthogonal Codes

A fiber optic channel must have the ability for multiple users to share the channel without interference. In order to facilitate this, optical orthogonal codes were developed by Salehi [70]. Viewing these codes as sets of integers modulo n leads to interesting design theoretic questions.

The study of optical orthogonal codes was first motivated by an application in a fiber optic code-division multiple access channel. Many users wish to transmit information over a common wide-band optical channel. The objective is to design a system that allows the users to share the common channel. Other approaches have included frequency division, time division, collision detection or some type of network synchronization. Each required frequent conversions between the optical domain and the electrical domain. However, employing a code-division multiple access system with optical orthogonal codes reduces the complexity of the system, enabling implementation with available technology and with potentially higher transmission efficiency [31].

An $(n, w, \lambda_a, \lambda_c)$ *optical orthogonal code* (OOC), C, is a family of $(0, 1)$- sequences of length n and weight w satisfying the following two properties (all subscripts are reduced modulo n):

1. $\sum_{0 \le t \le n-1} x_t x_{t+i} \le \lambda_a$ for any $\mathbf{x} = (x_0, x_1, \dots, x_{n-1})$ and any integer $i \not\equiv$ 0 mod n (the *auto-correlation property*);
2. $\sum_{0 \le t \le n-1} x_t y_{t+i} \le \lambda_c$ for any $\mathbf{x} = (x_0, \dots, x_{n-1})$, $\mathbf{y} = (y_0, \dots, y_{n-1})$, and any integer $i \not\equiv 0$ mod n (the *cross-correlation property*).

When $\lambda_a = \lambda_c = \lambda$ the code is an (n, w, λ) OOC. Research has concentrated on this case. To simplify the discussion of the model, let C be an $(n, w, 1)$ OOC with m codewords. The communications system can handle up to m simultaneous transmitters. Each transmitter is assigned one codeword from C, so that transmitter T_i is assigned codeword $\{s_1, s_2, \dots, s_w\} = c_i \in C$ (s_j indicates the position of the jth 1 in the $(0,1)$ sequence). At the transmitter, every information bit of a signal is encoded into a frame of n optical chips: If the information bit is 1, then in the corresponding frame (consisting of n optical chips), photon pulses are sent at exactly the s_1th, s_2th, $\dots$, s_wth chips. In the other $n - w$ chips, no photon pulses are sent. If, however, the information bit is 0, then no photon pulses are sent in the corresponding frame (still consisting of n chips). For example, if transmitter T_i wishes to send the message 101, this gets encoded as the sequence of 3 frames of length $3n$ where photon pulses are sent at times $s_1, s_2, \dots, s_w, 2n + s_1, 2n + s_2, \dots, 2n + s_w$.

All m users are allowed to transmit at any time; no network synchronization is required. At the receiving end, decoders separate the transmitted signals. The decoder consists of a bank of m tapped delay-lines, one for each codeword (so say D_i is the decoder for transmitter T_i). These delay taps on decoder D_i are a (possibly null) cyclic shift modulo n of those on T_i.

Each tapped delay-line can effectively calculate the correlation of the received waveform with its signature sequence. By the properties of OOCs, the correlation between different signature sequences is low. The delay-line output is high only when the intended transmitter's information bit is a 1. In particular, the output is w when decoder D_i receives the information bit 1 from transmitter T_i and they are synchronized correctly. The output is s for some $s \le m$ when D_i is not synchronized with T_i and T_i is sending a 1 or when T_i is sending a 0. Thus the receiver can effectively determine when the corresponding transmitter is transmitting an information bit of 1. "Bit stuffing" inserts a '1' in a prescribed manner after a specified number of consecutive '0's are transmitted, to ensure that the receiver can determine when long strings of 0's are sent.

OOCs are (0,1) sequences and are intended for environments that have no negative components. Most other correlation sequences are $(+1, -1)$ sequences intended for systems having both positive and negative components [31].

Research has also been done on using optical orthogonal codes for multimedia transmission in fiber-optic LANs [64] and in multirate fiber-optic CDMA systems [63]. The mathematical theory is quite similar to that described here.

A convenient way of viewing OOCs is from a set-theoretic perspective. An $(n, w, \lambda_a, \lambda_c)$ optical orthogonal code C can be considered as a family of w-sets of integers modulo n, in which each w-set corresponds to a codeword and the numbers in each w-set specify the nonzero bits of the codeword. The correlation properties can be rephrased in this set-theoretic framework. As an example, $C = \{1100100000000, 1010000100000\}$ is a (13,3,1) code with two codewords. In set theoretic notation, $C = \{\{0, 1, 4\}, \{0, 2, 7\}\}$ mod 13. The code is equivalent to a $(13, 3, 1)$ difference family in $\mathbb{Z}_{13}$ (which gives a Steiner triple system of order 13). Considering the set-theoretic interpretation of OOCs, it is to be expected that many of the constructions for OOCs are design theoretic in nature. This section outlines some constructions for OOCs that involve designs. To be consistent with design theoretic notation, we write $v = n$ and $k = w$, to speak of (v, k, λ) OOCs.

The main connection is to difference packings. Let $\mathcal{B} = \{B_1, B_2, \dots, B_t\}$, where $B_i = \{b_{i1}, b_{i2}, \dots b_{ik}\}, b_{ij} \in \mathbb{Z}_v, 1 \le i \le t$ and $1 \le j \le k$. The differences in $\mathcal{B}$ are $D = \{b_{ij} - b_{is} : 1 \le i \le t, 1 \le j, s \le k, j \ne s\}$. The pair $(\mathbb{Z}_v, \mathcal{B})$ is a *cyclic difference packing* or $CP(v, k, \lambda)$ if the cardinality of D is exactly $\lambda k(k - 1)t$ and $0 \notin D$. It can be easily verified that a (v, k, λ) cyclic difference packing gives a (v, k, λ) optical orthogonal code. A $CP(v, k, 1)$ is termed *g-regular* if the *difference leave* $(\mathbb{Z}_v \setminus D)$ along with 0 forms an additive subgroup of $\mathbb{Z}_v$ having order g. When $\lambda = 1$ a cyclic difference packing satisfies the bound $t \le \lfloor (v - 1)/(k(k - 1)) \rfloor$, and is *optimal* if $t = \lfloor (v - 1)/(k(k - 1)) \rfloor$.

A similar bound pertains to OOCs. Let $\Phi(v, k, \lambda)$ denote the maximum number of codewords in a (v, k, λ) OOC. Analogous to the Johnson bound from coding theory (see [62]), we have $\Phi(v, k, \lambda) \le ((v - 1)(v - 2) \cdots (v - \lambda))/(k(k - 1)(k - 2) \cdots (k - \lambda))$ [31].

When $\lambda = 1$ this reduces to $\Phi(v, k, 1) \le (v - 1)/(k(k - 1))$. When $|C| = \lfloor (v - 1)/(k(k - 1)) \rfloor$ the code is an *optimal OOC*. Moreover, the existence of an optimal $(v, k, 1)$ OOC is equivalent to the existence of an optimal $(v, k, 1)$ cyclic difference packing [87].

Hence results on (optimal) cyclic difference packings directly relate to results on (optimal) OOCs. When $k = 3$ and $\lambda = 1$, optimal OOCs arise from cyclic Steiner triple systems. In fact, when $v \equiv 1, 3 \pmod 6$ they coincide. Chung, Salehi and Wei [31] solved all cases when $v \not\equiv 2 \pmod 6$: $\Phi(v, 3, 1) = \lfloor \frac{(v-1)}{6} \rfloor$ if $v \not\equiv 2 \pmod 6$.

The proof is by a direct construction of a $(v, 3, 1)$ cyclic difference packing and is very reminiscent of the construction of cyclic triple systems from Skolem sequences (see [40]). In a similar vein, Yin [87] summarized results concerning optimal $(v, k, 1)$ cyclic difference packings, and hence gives many optimal OOCs. Most of these have $k = 4$, but there are some results for other $k \le 11$.

Another class of optimal OOCs comes from projective geometry. Chung, Salehi and Wei [31] used $\mathrm{PG}(d, q)$ to construct a $(v, k, 1)$ OOC where $v = (q^{d+1} - 1)/(q-1)$ and $k = q+1$. Let α be a primitive element of $\mathrm{GF}(q^{d+1})$ and say that $\log \beta = e$ if $\beta = \alpha^e$. Now, in the vector space $V(d+1, q)$ the nonzero vectors on a line ℓ through the origin are $\ell = \{\alpha^i, \alpha^{i+v}, \alpha^{i+2v}, \ldots, \alpha^{i+(q-2)v}\}$ where again $v = (q^{d+1} - 1)/(q-1)$. For any point $p \in \mathrm{PG}(d, q)$, let $\log p$ denote the log of any vector on the line corresponding to p in $V(d+1, q)$ modulo v. Hence each line in the projective geometry corresponds to a subset of the integers modulo v.

Let a *cyclic shift* of a line L in $\mathrm{PG}(d, q)$ be the set of points

$$\{p : \ \log p = 1 + \log p' \pmod{v} \text{ for some point } p' \in L\}.$$

The cyclic shift of a line is also a line in $\mathrm{PG}(d, q)$ so this creates a number of orbits of lines. If the number of lines in an orbit is v, then the orbit is termed *full*; otherwise it is a *short* orbit.

To construct an $(v, k, 1)$ OOC from the projective geometry, take one representative line from each full orbit and map each of these lines to the set of integers modulo v under the action of the log. These sets satisfy the autocorrelation and cross-correlation restrictions. The OOCs formed in this manner are optimal. Indeed, suppose that q is an odd prime power, $v = (q^{d+1} - 1)/(q-1)$ and $k = q+1$. Then $\Phi(v, k, 1) = \lfloor \frac{(v-1)}{k(k-1)} \rfloor$.

This construction can be extended to $\lambda > 1$ by using s-dimensional subspaces instead of 1-dimensional subspaces (lines).

A construction by Chen, Ge and Zhu [26] gives $(6v, 4, 1)$ optimal OOCs for infinitely many odd values of v. They construct these OOCs directly from skew starters. A *starter* in the cyclic group $\mathbb{Z}_v$ (v odd) is a set of unordered pairs $S = \{\{x_i, y_i\} : 1 \leq i \leq (v-1)/2\}$ which satisfies the two properties (1)$\{x_i : 1 \leq i \leq (v-1)/2\} \cup \{y_i \ 1 \leq i \leq (v-1)/2\} = \mathbb{Z}_v \backslash \{0\}$ and (2) $\{\pm(x_i - y_i) : 1 \leq i \leq (v-1)/2\} = \mathbb{Z}_v \backslash \{0\}$. The starter is a *skew starter* if in addition it satisfies the property (3) $\{\pm(x_i + y_i) : 1 \leq i \leq (v-1)/2\} = \mathbb{Z}_v \backslash \{0\}$. Skew starters have been useful in the construction of Room squares, Hamiltonian path balanced tournament designs and other combinatorial designs. See [47] for a survey.

Assume that $\gcd(v, 6) = 1$ and that $S = \{\{x_i, y_i\} : 1 \leq i \leq (v-1)/2\}$ is a skew starter in $\mathbb{Z}_v$. In $\mathbb{Z}_v \times \mathbb{Z}_6$ (which is isomorphic to $\mathbb{Z}_{6v}$), let $c_i = \{(x_i, 0), (y_i, 0), (x_i + y_i, 1), (0, 4)\}$ for $1 \leq i \leq (v-1)/2$. They show that the set $C = \{c_1, c_2, \ldots, c_{(v-1)/2}\}$ forms an $(6v, 4, 1)$ optimal OOC. Using results on the existence of skew starters they establish that there exists an optimal $(6v, 4, 1)$ OOC in $\mathbb{Z}_{6v}$ for all v such that $\gcd(v, 6) = 1$.

7   Superimposed Codes

This section examines a collection of applications to the design of codes for simultaneous communication, and to experimental design of pooling strategies. Du and Hwang [49] provide a much more detailed treatment.

A population $\mathcal{P}$ of b items contains a number d of *defective* items, and the remaining $b - d$ items are *good*. Items can be pooled together for testing: for a subset $X \subseteq \mathcal{P}$, the *group test* reports "yes" if X contains one or more defective elements, and reports "no" otherwise. The objective is to determine, using a number of group tests, precisely which items are defective. When group tests are all undertaken in parallel, the problem is *nonadaptive*; otherwise it is *adaptive*. Then results from one or more tests are available while constructing further pools to be tested. Among adaptive testing methods, some operate in a limited number of stages or rounds.

Group testing was first studied in screening large populations for disease [48], and with the advent of large-scale HIV screening, it has grown in importance. It has also arisen in satellite communications [11,83]. In this application, a large number of ground stations which rarely communicate share a satellite link. Rather than polling the ground stations individually, pools of the ground stations are formed as part of the system design. When the satellite enters a phase of accepting requests for reservations of time slots, it polls each pool and from the positive results on the pools it determines which ground stations wish to transmit. The satellite may have many positive responses within one pool, but detects only that there is at least one response. Hence, while cosmetically similar to the optical communication situation, this problem encounters unions rather than sums of colliding signals.

Let $\mathcal{P}$ be a set of b items, and let $\mathcal{X}$ be a collection of subsets of $\mathcal{P}$ corresponding to the group tests performed. Then $(\mathcal{P}, \mathcal{X})$ is a solution to the nonadaptive group testing problem if and only if, for any possible sets D_1 and D_2 of defective items, $\{X : D_1 \cap X \neq \emptyset, X \in \mathcal{X}\} = \{X : D_2 \cap X \neq \emptyset, X \in \mathcal{X}\}$ only if $D_1 = D_2$.

The dual of a solution $(\mathcal{P}, \mathcal{X})$ is a pair $(V, \mathcal{B})$, where the v group tests of $\mathcal{X}$ are in one-to-one correspondence with the points of V, and the b items are in correspondence with the blocks of $\mathcal{B}$ (for each item, the corresponding block contains the elements corresponding to the group tests containing the item). Typically $(V, \mathcal{B})$ is referred to as a solution to the group testing problem; the goal is to maximize the number of blocks (items tested) as a function of the number of points (group tests performed).

Often it is known with high probability that the number of defectives d does not exceed some threshold value p. In the *hypergeometric* problem, the number of defectives is assumed never to exceed p, and hence it is necessary that $(V, \mathcal{B})$ has the union of any two distinct sets, each containing at most p blocks, themselves distinct. In the *strict* problem, it is necessary to identify the set of defective items correctly when $d \leq p$ and to report when $d > p$. In the latter case, the specific set of defective items need not be determined, however.

Now consider a solution $(V, \mathcal{B})$ to the nonadaptive group testing problem with d defectives. Form a $|V| \times |\mathcal{B}|$ incidence matrix. This matrix has the property that the unions of two sets of at most d columns are distinct. The matrix is then $\bar{d}$-*separable* [49], and the corresponding set system is d-*union free* [51,54]. The columns of a $\bar{d}$-separable matrix form a superimposed code [50,60] which permits up to d simultaneously transmitted codewords to be unambiguously decoded. The decoding technique appears somewhat involved, because we could in principle be required to examine all unions of up to d columns. Hence a related family of matrices (or codes, or set systems) arises. If the incidence matrix contains no collection of d columns whose union covers a column not in the collection, then M is a d-*disjunct* matrix. If a disjunct matrix is employed, there is a simple decoding mechanism, observing that all codewords covered by the received union are 'positive'. Equivalently, we can alter the condition on the set system to require that it is d-*cover free*, i.e. that no union of d or fewer blocks contains another. Evidently, a d-cover free family is also d-union free.

Probabilistic bounds on the maximum numbers of blocks in cover free and union free families are available [49]; see [50,69,75] for bounds for cover free families. See [43] for progress in the union free case. Erdős, Frankl, and Füredi [51] established that among cover free families with constant block size, the maximum is realized by a Steiner t-design S($\ell, 2\ell - 1, m$); indeed Balding and Torney [7] recommend the use of an S(3,5,65) in a genetic application. For union free families, Frankl and Füredi [54] noted that Steiner triple systems give the largest 2-union free families when the block size is three; by permitting block size *at most* three, Vakil and Parnes [77] established a somewhat larger exact bound using group divisible designs with block size three.

In the *error correction* version of group testing, some group tests are permitted to report "false positives"; an *a priori* bound q on the number of such false positives is assumed. Balding and Torney [6] observed that $(V, \mathcal{B})$ is a solution to the strict group testing problem with threshold p and error correction for q false positives if and only if, for every union of p or fewer blocks, every other block contains at least $q + 1$ points not in this union. Any packing $(V, \mathcal{B})$ of t-sets into k-sets having $k \geq p(t - 1) + q + 1$ is a solution to the strict group testing problem with threshold p and error correction for q false positives. A Steiner system S($t, 2t - 1, v$) is a solution to the strict group testing problem with $p = 2$ and $q = 0$ that has the maximum number of blocks of any solution [6].

Finally, we consider the use of combinatorial designs in two-stage group testing. Here the objective in a first stage of pools is not to identify all defectives precisely, but rather to identify a small subset of the items which is guaranteed to contain all defective items. Frankl and Füredi call a family of sets d-*weakly union free* if, whenever two *disjoint* sets of blocks are chosen, each containing d or fewer blocks, their unions are distinct. A 2-weakly union free family with block size three provides pools for a group testing method for $d = 2$, in which a set of at most three potential defectives are identified [25]. Moreover, while union free families have no more blocks than a Steiner triple system has, weakly union free families can have twice as many blocks [54]. Chee, Colbourn, and Ling [25] established that certain twofold triple systems realize the bound. Not any twofold triple system forms a weakly union free family; four forbidden configurations of four blocks each must be avoided. Again, while the bound of Frankl and Füredi [54] suggests that designs can realize the maximum, the particular designs needed require additional structural properties [25]. Applications of designs in general in two-stage group testing appear to be just being explored; see [10] for useful observations.

Further applications of group testing arise in genomics [6,7,23], and in the construction of frameproof codes, which are designed to avoid coalitions of users forging the signature of a user not in the coalition [73,74].

8 Concluding Remarks

A general theme emerges from the applications described here. Restricting or balancing sizes of sets, numbers of subsets of various sizes that they contain, and intersection sizes among the subsets, often leads to 'equitable' allocation of resources to users or limits their interference. When this occurs, families of combinatorial designs arise naturally. This connection between applications and the theory of combinatorial designs can be seen to provide both useful techniques in the applications domain, and new and challenging problems in the combinatorial one.

Acknowledgments. The author's research is supported by ARO grant DAAG55-98-1-0272. Thanks to Myra Cohen, Jeff Dinitz, Alan Ling, Doug Stinson, and Paul van Oorschot for interesting discussions about this work.

References

[1] N. Alon, Expanders, sorting in rounds and superconcentrators of limited depth, *Proceedings of the Seventeenth ACM Symposium on the Theory of Computing* (1985) 98–102.

[2] C. Argon and H. Farooq Ahmad, Optimal optical orthogonal code design using difference sets and projective geometry, *Optics Communications* **118** (1995) 505–508.

[3] G. E. Atkin and H. P. Corrales, An efficient modulation/coding scheme for MFSK systems on bandwidth controlled channels, *IEEE Transactions on Sel. Areas Communications* **7** (1989) 1396–1401.

[4] G. E. Atkin, D. A. Fares and H. P. Corrales, Coded multipulse position modulation in a noisy optical channel, *Microwave Optical Technology Letters* **2** (1989) 336–340.

[5] B. E. Aupperle and J. F. Meyer, Fault-tolerant BIBD networks, *Proceedings of the Eighteenth International Symposium on Fault Tolerant Computing* IEEE, Piscataway NJ, 1988, pp. 306–311.

[6] D. J. Balding and D. C. Torney, Optimal pooling designs with error detection, *Journal of Combinatorial Theory (A)* **74** (1996) 131–140.

[7] D. J. Balding and D. C. Torney, The design of pooling experiments for screening a clone map, *Fungal Genetics and Biology* **21** (1997) 302–307.

[8] S. Ball, A. Blokhuis and F. Mazzocca, Maximal arcs in Desarguesian planes of odd order do not exist, *Combinatorica* **17** (1997) 31–41.

[9] A. Barlotti, Su $\{k - n\}$-archi di un piano lineare finito, *Boll. Un. Mat. Ital.* **11** (1956) 553–556.

[10] T. Berger and J. W. Mandell, Bounds on the efficiency of two-stage group testing, preprint, Cornell University, 1998.

[11] T. Berger, N. Mehravari, D. Towsley and J. Wolf, Random multiple-access communications and group testing, *IEEE Transactions on Communications* **32** (1984) 769–778.

[12] E. Berkovich and S. Berkovich, A combinatorial architecture for instruction-level parallelism, *Microprocessors and Microsystems* **22** (1998) 23–31.

[13] J. C. Bermond, J. Bond and S. Djelloul, Dense bus networks of diameter 2, *Interconnection Networks and Mapping and Scheduling Parallel Computations* (D. F. Hsu, A. L. Rosenberg and D. Sotteau; eds.) American Mathematics Society, Providence RI, 1994, pp. 9–16.

[14] J. C. Bermond, J. Bond, M. Paoli and C. Peyrat, Graphs and interconnection networks: diameter and vulnerability, *Surveys in Combinatorics 1983* (E. K. Lloyd; ed.) Cambridge University Press, Cambridge, 1983, pp. 1–30.

[15] J. C. Bermond, J. Bond and J. F. Saclé, Large hypergraphs of diameter 1, *Graph Theory and Combinatorics* (B. Bollobás; ed.) Academic Press, London, 1984, pp. 19–28.

[16] J. -C. Bermond and F. Ö. Ergincan, Bus interconnection networks, *Discrete Applied Mathematics* **68** (1996) 1–15.

[17] J.-C. Bermond, L. Gargano, S. Perennes, A. Rescigno, and U. Vaccaro, Efficient collective communication in all–optical networks, *Lecture Notes in Computer Science* 1099 (1996), 574–585.

[18] J.-C. Bermond, C. Huang, A. Rosa, and D. Sotteau, Decompositions of complete graphs into isomorphic subgraphs with five vertices, *Ars Combinatoria* 10 (1980), 211–254.

[19] Th. Beth, D. Jungnickel and H. Lenz, *Design Theory* Cambridge University Press, Cambridge, 1986.

[20] J. Bierbrauer, Small islands, *Extremal problems for finite sets (Visegrád, 1991)*, Bolyai Soc. Math. Stud., 3, János Bolyai Math. Soc., Budapest, 1994, pp. 93–110.

[21] C. M. Bird and A. D. Keedwell, Design and applications of optical orthogonal codes—a survey *Bulletin of the Institute for Combinatorics and its Applications* **11** (1994) 21–44.

[22] S. Bitan and T. Etzion, On constructions for optimal optical orthogonal codes, *Lecture Notes in Computer Science* **781** (1994) 111–125.

[23] W. J. Bruno, D. J. Balding, E. H. Knill, D. Bruce, C. Whittaker, N. Doggett, R. Stallings and D. C. Torney, Design of efficient pooling experiments, *Genomics* **26** (1995) 21–30.

[24] M. Buratti, A powerful method for constructing difference families and optimal optical orthogonal codes *Designs Codes and Cryptography* **5** (1995) 13–25.

[25] Y. M. Chee, C. J. Colbourn and A. C. H. Ling, Weakly union-free twofold triple systems, *Annals Combinatorics* **1** (1997) 215–225.

[26] K. Chen, G. Ge and L. Zhu, Starters and related codes, *Journal of Statistical Planning and Inference* **86** (2000), 379-395.

[27] A. L. Chiu and E. H. Modiano, Traffic grooming algorithms for reducing electronic multiplexing costs in WDM ring networks, *Journal of Lightwave Technology* **18** (2000) 2–12.

[28] F. R. K. Chung, Zone-balanced networks and block designs, *Bell System Technical Journal* **57** (1978) 2957–2981.

[29] F. R. K. Chung, On concentrators, superconcentrators, generalizers and non-blocking networks, *Bell System Technical Journal* **58** (1979) 1765–1777.

[30] F. R. K. Chung, On switching networks and block designs II, *Bell System Technical Journal* **59** (1980) 1165–1173.

[31] F. R. K. Chung, J. A. Salehi and and V. K. Wei, Optical orthogonal codes: design, analysis and applications, *IEEE Transactions on Information Theory* **35** (1989) 595–604. Correction: *IEEE Transactions on Information Theory* **38** (1992) 1429.

[32] M. B. Cohen and C. J. Colbourn, Steiner triple systems as multiple erasure correcting codes in disk arrays, *Proceedings of IPCCC 2000 (19th IEEE International Conference on Performance, Computing and Communications)*, 2000, pp. 288–294.

[33] M. B. Cohen and C. J. Colbourn, Optimal and pessimal orderings of Steiner triple systems in disk arrays, LATIN 2000 (Punta del Este, Uruguay), *Lecture Notes in Computer Science* 1776 (2000), 95-104.

[34] C. J. Colbourn, Weakly union-free maximum packings, *Annals Combinatorics* **3** (1999), 43–52.

[35] C. J. Colbourn, Projective planes and congestion-free networks, preprint, University of Vermont, 1998.

[36] C. J. Colbourn and J. H. Dinitz (editors), *CRC Handbook of Combinatorial Designs* CRC Press, Boca Raton FL, 1996.

[37] C. J. Colbourn, J. H. Dinitz, and D. R. Stinson, Applications of combinatorial designs to communications, cryptography, and networking, *Surveys in Combinatorics 1999* (J.D. Lamb and D.A. Preece; eds.), Cambridge University Press, pp. 37-100.

[38] C. J. Colbourn and A. C. H. Ling, Quorums from difference covers, *Information Processing Letters*, to appear.

[39] C. J. Colbourn and P. C. van Oorschot, Applications of combinatorial designs in computer science, *ACM Computing Surveys* **21** (1989) 223–250.

[40] C. J. Colbourn and A. Rosa, *Triple Systems* Oxford University Press, Oxford, 1999.

[41] C. J. Colbourn and S. Zhao, Maximum Kirkman signal sets for synchronous unipolar multi-user communication systems, *Designs, Codes and Cryptography* **20** (2000), 219–227.

[42] M.J. Colbourn and R.A. Mathon, On cyclic Steiner 2-designs, *Annals of Discrete Mathematics* **7** (1980), 215–253.

[43] D. Coppersmith and J. B. Shearer, New bounds for union-free families of sets, *Electron. Journal of Combinatorics* **5** (1998) #R39.

[44] W. W. M. Dai, Y. Kajitani and Y. Hirata, Optimal single hop multiple bus networks, *Proceedings of the 1993 IEEE International Symposium on Circuits and Systems* IEEE, Piscataway NJ, 1993, pp. 2541–2544.

[45] R. H. F. Denniston, Some maximal arcs in finite projective planes, *Journal of Combinatorial Theory* **6** (1969) 317–319.

[46] J. H. Dinitz and D. R. Stinson (editors), *Contemporary Design Theory: A Collection of Surveys* John Wiley & Sons, New York, 1992.

[47] J. H. Dinitz and D. R. Stinson, Room squares and related designs, in [46], 137–204.

[48] R. Dorfman, The detection of defective members of a large population, *Annals of Mathematical Statistics* **14** (1943) 436–440.

[49] D. Z. Du and F. K. Hwang, *Combinatorial Group Testing and Its Applications* World Scientific, Singapore, 1993.

[50] A. D'yachkov, V. Rykov and A. M. Rashad, Superimposed distance codes, *Problems Control and Information Theory* **18** (1989) 237–250.

[51] P. Erdős, P. Frankl and Z. Füredi, Families of finite sets in which no set is covered by the union of two others, *Journal of Combinatorial Theory (A)* **33** (1982) 158–166.

[52] D. A. Fares, Concatenated coding for multipulse signaling in noisy optical channels, *Microwave Optical Technology Letters* **4** (1991) 359–361.

[53] D. A. Fares, W. H. Abul-Shohoud, N. A. Raslan and M. A. Nassef, δmax detection of multipulse signaling in noisy optical channels, *Microwave Optical Technology Letters* **5** (1992) 269–273.

[54] P. Frankl and Z. Füredi, A new extremal property of Steiner triple systems, *Discrete Mathematics* **48** (1984) 205–212.

[55] Z. Füredi, Maximum degree and fractional matchings in uniform hypergraphs, *Combinatorica* **1** (1981) 155–162.

[56] A. Ghafoor, T. R. Bashkow and I. Ghafoor, Bisectional fault-tolerant communication architecture for supercomputer systems, *IEEE Transactions on Computers* **38** (1989) 1425–1446.

[57] J. W. P. Hirschfeld, *Projective Geometries Over Finite Fields, 2nd Ed.* Oxford University Press, Oxford, 1998.

[58] J. W. P. Hirschfeld and L. Storme, The packing problem in statistics, coding theory and finite projective spaces, *Journal of Statistical Planning and Inference* **72** (1998) 355–380.

[59] D. R. Hughes and F. C. Piper, *Projective Planes* Springer Verlag, New York, 1973.

[60] W. H. Kautz and R. R. Singleton, Nonrandom binary superimposed codes, *IEEE Transactions on Information Theory* **10** (1964) 363–377.

[61] J. Kilian, S. Kipnis and C. E. Leiserson, The organization of permutation architectures with bussed interconnections, *IEEE Transactions on Computers* **39** (1990) 1346–1358.

[62] F. J. MacWilliams and N. J. A. Sloane, *The Theory of Error-Correcting Codes* North-Holland, Amsterdam, 1978.

[63] S. V. Maric and V. K. N. Lau, Multirate fiber-optic CDMA: System design and performance analysis, *Journal of Lightwave Technology* **16** (1998) 9–17.

[64] S. V. Maric, O. Moreno and C. Corrada, Multimedia transmission in fiber-optic LANs using optical CDMA, *Journal of Lightwave Technology* **14** (1996) 2149–2153.

[65] M. D. Mickunas, Using projective geometry to design bus connection networks, *Proceedings of the Workshop on Interconnection Networks for Parallel and Distributed Processing* 1980, pp. 47–55.

[66] D. Raghavarao, *Constructions and Combinatorial Problems in Design of Experiments* John Wiley & Sons, New York, 1971.

[67] C.A. Rodger, Graph decompositions, *Le Matematiche* 45 (1990), 119-140.

[68] V. Rödl, On a packing and covering problem, *European J. Combin.* **6** (1985), 69–78.

[69] M. Ruszinkó, On the upper bound of the size of the r-cover-free families, *Journal of Combinatorial Theory (A)* **66** (1994) 302–310.

[70] J. A. Salehi, Code division multiple-access techniques in optical fibre networks – part I: Fundamental principles, *IEEE Transactions on Information Theory* **37** (1989) 824–833.

[71] J.M. Simmons, E.L. Goldstein, and A.A.M. Saleh, Quantifying the benefit of wavelength add–drop in WDM rings with distance-independent and dependent traffic, *J. Lightwave Technology* 17 (1999), 48–57.

[72] D. R. Stinson, Combinatorial designs and cryptography, *Surveys in Combinatorics, 1993* (K. Walker; ed.) Cambridge University Press, London, 1993, pp. 257–287.

[73] D. R. Stinson, Tran van Trung and R. Wei, Secure frameproof codes, key distribution patterns, group testing algorithms, and related structures, *Journal of Statistical Planning and Inference* **86** (2000), 595–617.

[74] D. R. Stinson and R. Wei, Combinatorial properties and constructions of traceability schemes and frameproof codes, *SIAM Journal of Discrete Mathematics* **11** (1998) 41–53.

[75] D. R. Stinson, R. Wei and L. Zhu, Some new bounds for cover-free families, *Journal of Combinatorial Theory (A)* **90** (2000) 224–234.

[76] L. Teirlinck, Non-trivial *t*-designs exist for all *t*, *Discrete Mathematics* **65** (1987), 345–356.

[77] F. Vakil and M. Parnes, On the structure of a class of sets useful in nonadaptive group testing, *Journal of Statistical Planning and Inference* **39** (1994) 57–69.

[78] P.-J. Wan, *Multichannel Optical Networks: Network Theory and Applications*, Kluwer Academic Press, 1999.

[79] D. Wiedemann, Cyclic difference covers through 133, *Congressus Numerantium* **90** (1992) 181–185.

[80] R.M. Wilson, An existence theory for pairwise balanced designs I: Composition theorems and morphisms, *Journal of Combinatorial Theory (A)* 13 (1971), 220–245.

[81] R.M. Wilson, An existence theory for pairwise balanced designs II. The structure of PBD-closed sets and the existence conjectures, *Journal of Combinatorial Theory (A)* 13 (1972), 246-273.

[82] R.M. Wilson, An existence theory for pairwise balanced designs III: A proof of the existence conjectures, *Journal of Combinatorial Theory (A)* 18 (1975), 71-79.

[83] J. K. Wolf, Born again group testing: multiaccess communications, *IEEE Transactions on Information Theory* **IT-31** (1985) 185–191.

[84] G. C. Yang, Some new families of optical orthogonal codes for code-division multiple-access fibre-optic networks, *IEEE Transactions on Communications* **142** (1995) 363–368.

[85] R. Yao, T. Chen and T. Kang, An investigation of multibus multiprocessor systems, *Acta Electron. Sinica* **18** (1990) 125–127.

[86] B. Yener, Y. Ofek and M. Yung, Combinatorial design of congestion-free networks, *IEEE/ACM Transactions on Networking* **5** (1997) 989–1000.

[87] J. X. Yin, Some combinatorial constructions for optical orthogonal codes, *Discrete Mathematics* **185** (1998) 201–219.

[88] S. Zhao, Application of BIBDs in MT-MFSK signal set design for multiplexing bursty sources, PhD thesis, University of Technology Sydney, 1998.

[89] S. Zhao, K. W. Yates and K. Yasukawa, Application of Kirkman designs in joint detection multiple access schemes, *Proceedings of the International Symposium on Spread Spectrum Techniques and Applications* **2** (1996) 857–861.

[90] S. Q. Zheng, Sparse hypernetworks based on Steiner triple systems, *Proceedings of the 1995 International Conference on Parallel Processing* IEEE, Piscataway NJ, 1995, pp. I.92–I.95.

[91] S. Q. Zheng, An abstract model for optical interconnection networks, *Parallel Computing Using Optical Interconnections* (K. Li, Y. Pan and S. Q. Zheng; eds.) Kluwer Academic, Norwell, MA, 1998, pp. 139–162.

Exact and Approximate Testing/Correcting of Algebraic Functions: A Survey

Marcos Kiwi[1] *, Frédéric Magniez[2] **, and Miklos Santha[2] ***

[1] Dept. Ing. Matemática, U. Chile & Ctr. Modelamiento Matemático, UMR 2071
UChile–CNRS, Santiago 170–3, Chile.
mkiwi@dim.uchile.cl
[2] CNRS–LRI, UMR 8623 Université Paris–Sud, 91405 Orsay, France.
{magniez,santha}@lri.fr

Abstract. In the late 80's Blum, Luby, Rubinfeld, Kannan et al. pioneered the theory of self–testing as an alternative way of dealing with the problem of software reliability. Over the last decade this theory played a crucial role in the construction of probabilistically checkable proofs and the derivation of hardness of approximation results. Applications in areas like computer vision, machine learning, and self–correcting programs were also established.

In the self–testing problem one is interested in determining (maybe probabilistically) whether a function to which one has oracle access satisfies a given property. We consider the problem of testing algebraic functions and survey over a decade of research in the area. Special emphasis is given to illustrate the scenario where the problem takes place and to the main techniques used in the analysis of tests. A novel aspect of this work is the separation it advocates between the mathematical and algorithmic issues that arise in the theory of self–testing.

1 Introduction

The issue of program (software) reliability is probably just as old as the theory of program design itself. People have spent and continue to spend considerable time on finding bugs in programs. But, the conception of a really satisfying theory for handling this problem remains a hard and elusive goal. Besides professional programmers, users would also like to dispose of tools which could enable them to efficiently address this task. Since they are usually not experts, these tools should ideally be less complicated than the ones used in the programs themselves. The fact that programs are becoming more and more involved obviously presents an

---

* Gratefully acknowledges the support of Conicyt via Fondecyt No. 1981182 and Fondap in Applied Mathematics, 2000.

** Partially supported by the EC thematic network RAND-APX IST-1999-14036. The participation at the Summer School was founded by the LRI (Orsay) and the IPM (Tehran).

*** Partially supported by the EC thematic network RAND-APX IST-1999-14036. The participation at the Summer School was founded by the EGIDE (Paris) and the IPM (Tehran).

additional difficulty. Nonetheless, several approaches have been considered and are used in practice. Each of them have different pros and cons. None is totally satisfactory.

The method of program verification proceeds through mathematical claims and proofs involving the behavior of a program. In principle this method could perfectly achieve the desired task once the program has been proven to behave correctly on all possible inputs. Another advantage of this approach is that the verification takes place only once, before the program is ever executed. Unfortunately, establishing such proofs turns out to be extremely difficult, and in practice it has only been achieved for a few quite simple programs. Also, requiring that programmers express their ideas in mathematically verifiable programs is probably not a realistic expectation. Moreover, there is no protection against errors caused by hardware problems.

Traditional program testing selects a few (sometimes random) inputs, and verifies the program's correctness on these instances. The drawbacks of this approach are fairly obvious. First, there is a priori no reason that the correctness on the chosen instances would imply correctness on instances which were not tested. Second, testing the correctness on the chosen instances usually involves another program which is believed to execute perfectly its task. Clearly, there is some circularity in this reasoning: relying on the correctness of another program is using a tool which is just as powerful as the task that was set to be achieved. Finally, hardware based errors might not be detected until it is too late.

In the late eighties a significantly novel approach, the theory of *program checking* and *self-testing/correcting*, was pioneered by the work of Blum [Blu88], Blum and Kannan [BK89] and Blum, Luby, and Rubinfeld [BLR90]. This theory is meant to address different aspects of the basic problem of program correctness via formal methods by verifying carefully chosen mathematical relationships between the outputs of the program on randomly selected inputs. Specifically, consider the situation where a program P is supposed to compute some function f. A *checker* for f verifies whether the program P computes f on a particular input x; a *self-tester* for f verifies whether the program P is correct on most inputs; and a *self-corrector* for f uses a program P, which is correct on most inputs, to compute f correctly everywhere. All these tasks are supposed to be achieved algorithmically by probabilistic procedures, and the stated requirements should be obtained with high probability. Checkers and self-testers/correctors can only access the program as a black box, and should do something different and simpler than to actually compute the function f.

More than a decade after the birth of this new approach, one can state with relatively high assurance that it has met considerable success both in the theoretical level and in practice. The existence of efficient self-testers was established in the early years of the theory for a variety of mostly algebraic problems, including linear functions and polynomials. These results which were first obtained in the model of exact computations were later partly generalized to more and more complicated (and realistic) models of computations with errors. Self-testers were heavily used in structural complexity, paving the way for the fundamental results characterizing complexity classes via interactive and probabilistically checkable proofs. These results also had remarkable and surprising consequences — they played a crucial role in the derivation of strong non–approximability results for NP–hard optimization problems. In recent years, the theory of self–testing has evolved into what is called today *property testing*, where one has to establish via a few random checks whether an object possesses some given property. Among the many examples one can mention numerous graph properties such as bipartiteness or colorability; monotonicity of functions, or properties of formal languages.

On the practical side, self–testers/correctors were constructed for example for a library of programs computing standard functions in linear algebra [BLR90]. The viability of the approach was also illustrated in the study by Blum and Wassermann [BW97] of the division bug of the first Pentium processors. They showed that this problem could have been detected and corrected by the self–testing/correcting techniques already available at that time. Also, the self–tester of Ergün [Erg95] for the Discrete Fourier Transform is currently used in the software package FFTW for computing reliably fast Fourier transformations [FFT].

In this survey we review the most important results and techniques that arise in self–testing/correcting algebraic functions, but we do not address the subject of checking since the existence of a self–tester/corrector directly implies the existence of a checker. This work also contains some new results about self–correcting, but its main originality lies in the systematic separation it advocates between the purely mathematical and the algorithmic/computational aspects that arise in the theory of self–testing. Also, we do not include any specific computational restriction in our definitions of self–testers/correctors. Instead, we think it is better to give precise statements about the algorithmic performance of the self–testers/correctors constructed. The advantage of this approach is that it allows to independently address different aspects of self–testers/correctors.

This work will be divided into two main parts: the first one deals with exact, and the second with approximate computations. In both models, our basic definition will be for testing function families. In the exact model we first prove a generic theorem for constructing self–testers. This method requires that the family to be tested be characterized by a (property) test possessing two specific properties: continuity and robustness. These properties ensure that the distance of a program from the target function family is close to the probability that the program is rejected by the test, which in turn can be well approximated by standard sampling techniques. After illustrating the method on the benchmark problem of linearity, we address the questions of self–correcting linear functions, and a way to handle the so-called generator bottleneck problem which is often encountered when testing whether a program computes a specific function. Afterwards, we study self–testers for multiplication and polynomials.

The basic treatment of the approximate model will be analogous. The general notion of a computational error term will enable us to carry this out in several models of approximate computing, such as computations with absolute error, with error dependent on the input size, and finally with relative error.

We will emphasize the new concepts and techniques we employ to deal with the increasing difficulties due to the changes in the model. In particular, we will have to address a new issue that arises in approximate testing: stability. This property ensures that a program approximately satisfying a test everywhere is close to a function which exactly satisfies the test everywhere. We formalize here the notion of an approximate self–corrector. In our discussion of approximate self–testers we again address the linearity testing problem in (almost) full detail, whereas for polynomials we mostly simply state the known results. On the other hand, we address in some detail the issue of how to evaluate rapidly the test errors in the case of input size dependent errors. In the case of relative error we discuss why the standard linearity test, which works marvelously well in all previously mentioned scenarios, has to be replaced by a different one.

In the last section of this work we briefly deal with two subjects closely related to self–testing: probabilistically checkable proofs and property testing. Probabilistically checkable proofs heavily use specific self–testing techniques to verify with as few queries as possible whether a function satisfies some pre-specified properties. Property testing applies the testing paradigm to the verification of properties of combinatorial objects like graphs, languages, etc. We conclude this survey by describing the relation between self–testing and property testing and mentioning some recent developments concerning this latter framework.

2   Exact Self-Testing

2.1   Introduction to the Model

Throughout this section, let D and R be two sets such that D is finite, and let C be a family of functions from D to R. In the testing problem one is interested in determining, maybe probabilistically, how "close" an oracle function $f : D \to R$ is from an underlying family of functions of interest $\mathcal{F} \subseteq C$. The function class $\mathcal{F}$ represents a property which one desires f to have. In order to formalize the notion of "closeness," the concept of distance is introduced. Informally, the distance between f and $\mathcal{F}$ is the smallest fraction of values taken by f that need to be changed in order to obtain a function in $\mathcal{F}$. For a formal definition, let $\Pr_{a \in_{\mathcal{D}} A} [E_a]$ denote the probability that the event E_a occurs when a is chosen at random according to the distribution $\mathcal{D}$ over A (typically, $\mathcal{D}$ is omitted when it is the uniform distribution).

Definition 1 (Distance). *Let $P, f \in C$ be two functions. The* distance between P *and* f *is*

$$\mathsf{Dist}(P, f) = \Pr_{x \in D} [P(x) \neq f(x)] .$$

If $\mathcal{F} \subseteq C$, then the distance of P from $\mathcal{F}$ *is*

$$\mathsf{Dist}(P, \mathcal{F}) = \mathop{\mathsf{Inf}}_{f \in \mathcal{F}} \mathsf{Dist}(P, f).$$

A self–tester for a function class $\mathcal{F} \subseteq C$ is a probabilistic oracle program T that can call as a subroutine another program $P \in C$, i.e., can pass to P an input x

and is returned in one step $P(x)$. The goal of T is to ascertain whether P is either close or far away from $\mathcal{F}$. This notion was first formalized in [BLR90].

Definition 2 (Self–tester). *Let $\mathcal{F} \subseteq \mathcal{C}$, and let $0 \leq \eta \leq \eta' < 1$. An (η, η')– self–tester for $\mathcal{F}$ on $\mathcal{C}$ is a probabilistic oracle Turing machine T such that for every $P \in \mathcal{C}$ and for every confidence parameter $0 < \gamma < 1$:*

- *if $\mathrm{Dist}(P, \mathcal{F}) \leq \eta$, then $\Pr\left[T^P(\gamma) = \texttt{GOOD}\right] \geq 1 - \gamma$;*
- *if $\mathrm{Dist}(P, \mathcal{F}) > \eta'$, then $\Pr\left[T^P(\gamma) = \texttt{BAD}\right] \geq 1 - \gamma$,*

where the probabilities are taken over the coin tosses of T.

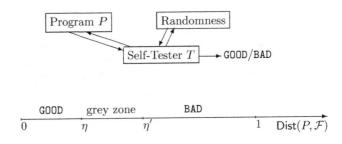

Fig. 1. Definition of an (η, η')–self–tester.

One of the motivations for building self–testers is to make it possible to gain evidence that a program correctly computes a function f on a collection of instances without trying to prove that the program is correct on all possible inputs. However, this raises the question of how to determine that the self–tester is correct. One way around this issue is to ask for the self–tester to be simpler than any correct program for f. Unfortunately simplicity is an aesthetic notion difficult to quantify. Thus, Blum suggested forcing the self–tester to be different from any program computing f in a quantifiable way. This leads to the following definition [Rub90]: *A self–tester T is quantifiably different with respect to $\mathcal{F}$, when for all programs P the incremental time taken by T^P is smaller than the fastest known program for computing a function in $\mathcal{F}$.*[1] Still, this requires the design of equally good self–testers for both efficient and inefficient programs purportedly computing the same function f. Moreover, self–testers are useful in contexts other than program verification, e.g., in the construction of probabilistically checkable proofs where one is more concerned with the query complexity and randomness usage rather than the efficiency of the self–testers.

---

[1] Ideally, one would prefer that the incremental time be smaller than any correct program for computing functions in $\mathcal{F}$. But, this is too strong a requirement since for many problems of interest no non–trivial lower bound on a correct program's running time is known.

Thus, we simply advocate precisely stating the incremental running time and the operations carried out by the self–testers in order to let the user judge whether the self–tester is useful.

Traditionally, the self–testing literature identifies a test with a self–tester. We do not advocate this practice. We prefer to think of a test as a purely mathematical object and keep it separate from its computational implementation, as proposed in [Kiw96]. This motivates the following:

Definition 3 (Exact test). *An* exact test $(\mathcal{T}, \mathcal{C}, \mathcal{D})$ *is a set of applications* $\mathcal{T}$ *from* $\mathcal{C}$ *to the set* {GOOD,BAD} *together with a distribution* $\mathcal{D}$ *over* $\mathcal{T}$. *The exact test characterizes the family of functions*

$$\mathsf{char}(\mathcal{T}, \mathcal{C}, \mathcal{D}) = \{f \in \mathcal{C} \ : \ \Pr_{t \in_{\mathcal{D}} \mathcal{T}} [t(f) = \text{GOOD}] = 1\}.$$

The rejection probability *of a function* $P \in \mathcal{C}$ *by the exact test is defined as*

$$\mathsf{Rej}(P, \mathcal{T}) = \Pr_{t \in_{\mathcal{D}} \mathcal{T}} [t(P) = \text{BAD}].$$

A probabilistic oracle Turing machine M *realizes the exact test* $\mathcal{T}$ *on* $\mathcal{C}$ *if for all* $P \in \mathcal{C}$,

$$\Pr \left[M^P \ \text{returns BAD} \right] = \mathsf{Rej}(P, \mathcal{T}),$$

where the probability on the left hand side is taken over the coin tosses of the machine M.

For the sake of clarity, we specify an exact test via the following mathematically equivalent very high level algorithm:

Exact Test$(P \in \mathcal{C}, \mathcal{T}, \mathcal{D})$
1. Choose an element $t \in \mathcal{T}$ according to $\mathcal{D}$.
2. Reject if $t(P) = \text{BAD}$ (otherwise accept).

This notation highlights how to realize an exact test: First, randomly sample from $\mathcal{T}$ according to $\mathcal{D}$ by using only coin tosses and then compute $t(P)$.

In order not to unnecessarily clutter the notation, when referring to an exact test $(\mathcal{T}, \mathcal{C}, \mathcal{D})$ we henceforth omit $\mathcal{D}$ and assume that it is the uniform distribution over $\mathcal{T}$. Also, if no reference to a particular distribution is given, by a randomly chosen element we mean an element chosen uniformly at random. In addition, when talking about several randomly chosen elements, unless said otherwise, we mean that they are randomly and independently chosen. It is a simple exercise to extend the framework presented here to the case of non–uniform distributions over $\mathcal{T}$. Note however, that if an exact test $\mathcal{T}$ on $\mathcal{C}$ is finite, then in the uniform distribution case it characterizes the family of functions $f \in \mathcal{C}$ such that $t(f) = \text{GOOD}$ for every $t \in \mathcal{T}$. We also omit $\mathcal{C}$ if it is clear from context. Finally, it is to be understood that a test accepts when it does not reject.

Computing the distance of a function from a family $\mathcal{F}$ is usually a hard task. On the other hand, the rejection probability of a function by an exact test $\mathcal{T}$ can be easily approximated by standard sampling techniques. Therefore, if an exact test characterizing some function family is such that for every function the rejection probability is close to the distance, then by approximating the rejection probability one can estimate the distance. This allows to probabilistically determine whether the oracle function is close or far away from the function class $\mathcal{F}$ of interest. In other words, one obtains a self–tester for $\mathcal{F}$. The two important properties of an exact test which ensure that this approach succeeds are:

Definition 4 (Continuity & robustness). *Let $\mathcal{T}$ be an exact test on $\mathcal{C}$ characterizing $\mathcal{F}$. Let $0 \leq \eta, \delta < 1$ be constants. Then $\mathcal{T}$ is (η, δ)–continuous if for all $P \in \mathcal{C}$,*

$$\mathsf{Dist}(P, \mathcal{F}) \leq \eta \Longrightarrow \mathsf{Rej}(P, \mathcal{T}) \leq \delta,$$

and it is (η, δ)–robust if for all $P \in \mathcal{C}$,

$$\mathsf{Rej}(P, \mathcal{T}) \leq \delta \Longrightarrow \mathsf{Dist}(P, \mathcal{F}) \leq \eta.$$

Thus, proving continuity of an exact test implies upper bounding the rejection probability of the exact test in terms of the relevant distance. On the contrary, to prove robustness one needs to bound the relevant distance in terms of the rejection probability of the exact test. In fact, we advocate explicitly stating these bounds as long as the clarity of the writeup is not compromised.

The importance of continuity and robustness was very early recognized in the self–testing literature. Proving continuity is usually very easy, often people do not even bother stating it explicitly. The term itself was first used by Magniez [Mag00a]. Robustness on the other hand is quite delicate to establish. The term itself was coined and formally defined by Rubinfeld and Sudan in [RS92b] and studied in [Rub94]. Typically, exact tests that are both continuous and robust give rise to self–testers. We now precisely state this claim. The construction of most of the known self–testers are based on it.

Theorem 1 (Generic self–tester). *Let $\mathcal{F} \subseteq \mathcal{C}$ be a function family and let $\mathcal{T}$ be an exact test on $\mathcal{C}$ that is realized by a probabilistic Turing machine M. Let $0 \leq \delta < \delta' < 1$ and $0 \leq \eta \leq \eta' \leq 1$. If $\mathcal{T}$ characterizes $\mathcal{F}$,*

- *is (η, δ)–continuous, and*
- *(η', δ')–robust,*

then there exists an (η, η')–self–tester T for $\mathcal{F}$ on $\mathcal{C}$. Moreover, T performs, for every confidence parameter $0 < \gamma < 1$, $O(\ln(1/\gamma)\frac{\delta + \delta'}{(\delta - \delta')^2})$ iterations of M, counter increments, comparisons, and binary shifts.

Proof. Let $\delta^* = (\delta + \delta')/2$. The self–tester T repeats N-times the computation of M with program P as oracle. After N repetitions, T computes the fraction err of runs which gave the answer BAD. If $err > \delta^*$, then T returns BAD, and GOOD otherwise. To make the computation of err simple, N is chosen a power of 2. Moreover, N is chosen large enough so that $\mathsf{Rej}(P, \mathcal{T}) \leq \delta$ (respectively $\mathsf{Rej}(P, \mathcal{T}) > \delta'$) implies $err \leq \delta^*$ (respectively $err > \delta^*$) with probability at least $1 - \gamma$. Standard Chernoff bound arguments (see e.g. [AS92a, Appendix A]) show that it is sufficient to choose N so that $N = O(\ln(1/\gamma)/\frac{\delta+\delta'}{(\delta-\delta')^2})$. The work performed by the self–tester consists of at most N iterations of M, counter increments, comparisons, and binary shifts.

We now show that T has the claimed properties. First, assume $\mathsf{Dist}(P,\mathcal{F}) \leq \eta$. The continuity of the exact test implies that $\mathsf{Rej}(P,\mathcal{T}) \leq \delta$. Therefore, with probability at least $1 - \gamma$ the machine $T^P(\gamma)$ computes $err \leq \delta^*$ and returns GOOD. Suppose now that $\mathsf{Dist}(P,\mathcal{F}) > \eta'$. Since the exact test is (η',δ')–robust, $\mathsf{Rej}(P,\mathcal{T}) > \delta'$. Therefore, with probability at least $1 - \gamma$ the machine $T^P(\gamma)$ computes $err > \delta^*$ and returns BAD.

A typical way of specifying an exact test is through a functional equation. Indeed, let $\Phi : \mathcal{C} \times \mathcal{N} \to \mathbb{R}$ be a functional where $\mathcal{N} \subseteq \bigcup_{k=1}^{|D|} D^k$. The set $\mathcal{N}$ is called the neighborhood set and each of its members is typically referred to as a neighborhood. The functional Φ induces the exact test $\mathcal{T}$ on $\mathcal{C}$ by defining for every $(x_1,\dots,x_k) \in \mathcal{N}$ the mapping $t_{x_1,\dots,x_k} : \mathcal{C} \to \{\text{GOOD}, \text{BAD}\}$ as

$$t_{x_1,\dots,x_k}(P) = \begin{cases} \text{GOOD,} & \text{if } \Phi(P,x_1,\dots,x_k) = 0, \\ \text{BAD,} & \text{otherwise,} \end{cases}$$

and letting

$$\mathcal{T} = \{t_{x_1,\dots,x_k} : (x_1,\dots,x_k) \in \mathcal{N}\}.$$

The **Exact Test**(P) thus becomes:

Functional Equation Test(P, Φ)
1. Randomly choose $(x_1,\dots,x_k) \in \mathcal{N}$.
2. Reject if $\Phi(P,x_1,\dots,x_k) = \text{BAD}$.

The family of functions characterized by the induced exact test consists of those functions $f \in \mathcal{C}$ satisfying the following functional equation:

$$\forall(x_1,\dots,x_k) \in \mathcal{N}, \qquad \Phi(f,x_1,\dots,x_k) = 0.$$

There might be different functionals characterizing the same collection of functions, and not necessarily all of them give rise to equally appealing exact tests. Indeed, one usually desires that the largest number of values of f that one needs to know in order to compute $\Phi(f,\dots)$, no matter what f is, be small. If the largest such number is K, the exact test is called K–local. For example, the exact test induced by the functional $\Phi(f,x,y) = f(x+y) - f(x) - f(y)$ is 3–local.

Through Theorem 1, functional equations that give rise to exact tests that are both continuous and robust lead to the construction of self–testers.

For the sake of concreteness, we now introduce one of the most famous self–testing problems, one that has become the benchmark throughout the self–testing literature for trying out new techniques, disproving conjectures, etc. — the so called *linearity testing problem*. In it, one is interested in verifying whether a function P taking values from one finite abelian group G into another such group G' is a group homomorphism. In other words, whether

$$\forall g, g' \in G, \qquad P(g+g') - P(g) - P(g') = 0.$$

This functional equation gives rise to the following functional equation test:

> **Linearity Test**(P)
> 1. Randomly choose $x, y \in G$.
> 2. Reject if $P(x + y) - P(x) - P(y) \neq 0$.

The above described exact test was introduced in [BLR90] and is also known as the BLR test. We will now illustrate the concepts introduced so far as well as discuss several important issues that arise in connection to testing by focusing our attention in the study of the **Linearity Test**.

2.2   Linearity Self-Testing

Let $\mathcal{C}$ denote the collection of functions from G to G', and let $\mathcal{L}$ be the subset of those functions that are homomorphisms. By Theorem 1, in order to come up with a self–tester for $\mathcal{L}$ on $\mathcal{C}$ we need only that the **Linearity Test** is both continuous and robust. As mentioned before, the continuity of an exact test is a property which is rather easy to establish. This is also the case for the **Linearity Test** as shown by the following result from which the $(\eta, 3\eta)$–continuity immediately follows.

Theorem 2. *Let G and G' be two finite abelian groups and let $P, g : G \to G'$ be such that g is a homomorphism. Then*

$$\Pr_{x,y \in G} [P(x + y) - P(x) - P(y) \neq 0] \leq 3\Pr_{x \in G} [g(x) \neq P(x)].$$

Proof. Observe that $P(x+y) - P(x) - P(y) \neq 0$ implies that $g(x+y) \neq P(x+y)$ or $g(x) \neq P(x)$ or $g(y) \neq P(y)$ and that $x + y$ is uniformly distributed in G for x, y uniformly and independently chosen in G. To conclude, apply the union bound. $\qed$

There is nothing very special about the **Linearity Test** for the above argument to work. Indeed, suppose that $\Phi(f, \dots)$ was a functional that gave rise to a K–local functional equation test. Then, the **Functional Equation Test** associated with Φ would be $(\eta, K\eta)$–continuous provided each evaluation of f was performed on an element chosen uniformly from a fixed subset of f's domain.

We now turn our attention to the harder task of proving robustness. In doing so we illustrate the most successful argument known for establishing this property — the so called majority function argument [BLR90,Cop89]. All proofs of robustness based on the majority function argument start by defining a function g whose value at x takes the most commonly occurring value among the members of a multiset S_x whose elements depend on x and P, i.e.,

$$g(x) = \mathsf{Maj}_{s \in S_x} (s).$$

(Here, as well as throughout this paper, $\mathsf{Maj}_{s \in S}(s)$ denotes the most frequent among the elements of the multiset S, ties broken arbitrarily.) Moreover, there are three clearly identified stages in this type of proof argument. First, one shows that an overwhelming number of the elements of each S_x agree with the most commonly occurring value in the set, i.e., $g(x)$. Second, it is shown that g is close to P. Finally, it is shown that g has the property of interest (in the case of Theorem 3, that g is an homomorphism). The majority argument as well as its three main stages are illustrated by the following result taken from [BLR90].

Theorem 3. *Let G and G' be two finite abelian groups and let $P : G \to G'$ be an application such that for some constant $\eta < 1/6$,*

$$\Pr_{x,y \in G} [P(x + y) - P(x) - P(y) \neq 0] \leq \eta.$$

Then, there exists a homomorphism $g : G \to G'$ such that

$$\Pr_{x \in G} [g(x) \neq P(x)] \leq 2\eta.$$

Proof. We define the function $g(x) = \mathsf{Maj}_{y \in G} (P(x + y) - P(y))$. First, we show that with overwhelming probability $P(c + y) - P(y)$ agrees with $g(c)$, i.e.,

$$\Pr_{y \in G} [g(c) = P(c + y) - P(y)] \geq 1 - 2\eta. \tag{1}$$

By hypothesis, for randomly chosen x and y in G, we have that $P(c + x + y) - P(c + x) - P(y) \neq 0$ with probability at most η. Under the same conditions, the probability that $P(c + x + y) - P(c + y) - P(x) \neq 0$ is also upper bounded by η. Therefore,

$$\Pr_{x,y \in G} [P(c + x) - P(x) = P(c + y) - P(y)] \geq 1 - 2\eta.$$

Note that $\sum_{z \in G'} (\Pr_{y \in G} [P(c + y) - P(y) = z])^2$ equals the left hand side term of the previous inequality. By definition of $g(c)$ we know that for every $z \in G'$

$$\Pr_{y \in G} [P(c + y) - P(y) = z] \leq \Pr_{y \in G} [g(c) = P(c + y) - P(y)].$$

Since $\sum_{z \in G'} \Pr_{y \in G} [P(c + y) - P(y) = z] = 1$, we obtain (1).

Suppose now, for the sake of contradiction, that the distance between P and g was greater than 2η. By (1), for every x the probability that $g(x) = P(x + y) - P(y)$ is at least $1/2$ when y is randomly chosen in G. Thus,

$$\Pr_{x,y \in G} [P(x + y) - P(x) - P(y) \neq 0] > \eta,$$

which contradicts our hypothesis.

Finally, we prove that g is indeed a homomorphism. Fix $a, b \in G$. Applying (1) three times we get that, with probability at least $1 - 6\eta$ when y is randomly chosen in G, the following three events hold

$$g(a) = P(a + y) - P(y),$$
$$g(b) = P(a + b + y) - P(a + y),$$
$$g(a + b) = P(a + b + y) - P(y).$$

Therefore,

$$\Pr_{y \in G} [g(a + b) = g(a) + g(b)] > 1 - 6\eta > 0.$$

Since the event $g(a + b) = g(a) + g(b)$ is independent of y, we get that $g(a + b) = g(a) + g(b)$ must hold.

Note that the proof of the previous result shows more than what its statement claims. In fact, the proof is constructive and it not only shows that an homomorphism g with the claimed properties exist, but that one such homomorphism is

$$g(x) = \mathsf{Maj}_{y \in G} \left(P(x + y) - P(y) \right).$$

Also, observe that a direct consequence of Theorem 3 is that the **Linearity Test** is $(2\eta, \eta)$–robust provided $\eta < 1/6$, or simply $(6\eta, \eta)$–robust (for every $\eta \geq 0$) if one is not so much concerned with the constants. We will use the latter statement since, rather than derive the best possible constants, in this work we strive to present ideas as clearly as possible. A similar convention is adopted throughout this survey for all the tests we discuss.

Corollary 1. *Let G and G' be two abelian groups, let C be the family of all functions from G to G', and let $\mathcal{L} \subseteq C$ be the set of homomorphisms. Then, for every $\eta > 0$, there is an $(\eta, 19\eta)$–self–tester for $\mathcal{L}$ on C which uses for every confidence parameter $0 < \gamma < 1$, $O(\ln(1/\gamma)/\eta)$ calls to the oracle program, additions comparisons, counter increments, and binary shifts.*

Proof. The **Linearity Test** characterizes $\mathcal{L}$ since it is induced by the functional equation

$$\forall x, y \in G, \qquad P(x + y) - P(x) - P(y) = 0.$$

Realizing the **Linearity Test** just means randomly choosing x and y in G and verifying whether $P(x + y) - P(x) - P(y) = 0$. By Theorem 2, the test is $(\eta, 3\eta)$–continuous and by Theorem 3 it is also $(6\eta', \eta')$–robust. Letting $\eta' = (3 + 1/6)\eta = 19\eta/6$ and applying Theorem 1, the existence of the claimed self–tester is established.

2.3   Self-Correcting

We saw that for some function classes $\mathcal{F}$, a self–tester can be used to ascertain whether a program P correctly computes a function in $\mathcal{F}$. As we shall see later on, self–testing techniques can often be used to verify (probabilistically) whether a program P computes a specific function $g \in \mathcal{F}$ on some significant fraction of its domain. Sometimes in these cases, the program P itself can be used to compute g correctly with very large probability *everywhere* in its domain. This leads to the following:

Definition 5 (Self–corrector). *Let $\mathcal{F} \subseteq C$ be a function family and let $\eta \geq 0$. An η–self–corrector for $\mathcal{F}$ on C is a probabilistic oracle Turing machine T such that for every $P \in C$, if $\mathsf{Dist}(P, g) \leq \eta$ for some $g \in \mathcal{F}$, then for every $x \in D$ and for every confidence parameter $0 < \gamma < 1$, the output $T^P(x, \gamma)$ is $g(x)$ with probability at least $1 - \gamma$.*

Note that by definition, in order to possess an η-self–corrector, a family $\mathcal{F}$ has to satisfy that for each function $P \in C$, there exists at most one function $g \in \mathcal{F}$ such that $\mathsf{Dist}(P, g) \leq \eta$.

Below we give an example of a self–corrector for the class of homomorphisms from one finite abelian group into another. In doing so we illustrate how the majority function argument discussed in the previous section naturally gives rise, when applicable, to self–correctors.

Theorem 4. *Let G and G' be two abelian groups, let C be the family of all functions from G to G', and let $\mathcal{L} \subseteq \mathcal{C}$ be the set of homomorphisms. Then, for every $0 \leq \eta < 1/4$ there is an η–self–corrector for $\mathcal{L}$ on $\mathcal{C}$ which uses for every confidence parameter $0 < \gamma < 1$, $O(\ln(1/\gamma)/(1 - 4\eta)^2)$ calls to the oracle program, additions, comparisons, counter increments, and binary shifts.*

Proof. For some fixed $P \in \mathcal{C}$, let g be the function defined in $x \in G$ by $g(x) = \mathsf{Maj}_{y \in G} \left(P(x + y) - P(y) \right)$.

If $\mathsf{Dist}(P, \mathcal{L}) \leq \eta$ for some $0 \leq \eta < 1/4$, then there exists only one function $l \in \mathcal{L}$ such that $\mathsf{Dist}(P, l) \leq \eta$. This is a direct consequence of the fact that for all linear functions $l, l' \in \mathcal{L}$, since the elements $x \in G$ such that $l(x) = l'(x)$ form a subgroup H of G, either $H = G$ or $|H|/|G| \leq 1/2$. Therefore, either $l = l'$ or $\mathsf{Dist}(l, l') \geq 1/2$.

The closeness of P to l implies that $P(x + y) - P(y) = l(x)$ for at least half the elements y in G. Thus, by the definition of g, we get that $g = l$. Moreover, Chernoff bounds tell us that for every $x \in G$, the quantity $g(x)$ can be correctly determined with probability greater than $1 - \gamma$ by choosing $N = O(\ln(1/\gamma)/(1 - 4\eta)^2)$ points $y_i \in G$ and then computing $\mathsf{Maj}_{i=1,\dots,N} \left(P(x + y_i) - P(y_i) \right)$.

Note that the function $g(x) = \mathsf{Maj}_{y \in G} \left(P(x + y) - P(y) \right)$ played a key role both in the construction of a self–tester and of a self–corrector for the function class of homomorphisms. This is a distinctive feature of the majority function argument. Indeed, recall that this argument is constructive. Specifically, it proceeds by defining a function g whose value at x takes the most commonly occurring value among the members of a set S_x whose elements depend on x and P. When successful, this argument suggests that a self–corrector for the function class of interest is one that on input γ and x, for an appropriately chosen $N = N(\gamma)$, randomly chooses $g_1, \dots, g_N$ in S_x and returns $\mathsf{Maj}_{i=1,\dots,N} (g_i)$.

2.4   Generator Test

An extreme case of self–testing is to ascertain whether a program P computes a fixed function f. This task is usually undertaken in two stages. First, self–testing techniques are used in order to determine whether P computes, in a large fraction of its inputs, a function f_P in some specific function class $\mathcal{F}$. For example, when f is a group homomorphism, one checks whether P is close to a group homomorphism. In the second stage, one ascertains whether f_P is indeed equal to f. To do so it suffices to check that f_P and f agree in a collection of inputs equality over which implies agreement everywhere. The process through which this goal is achieved is called *generator test*. This is somewhat complicated by the fact that one cannot evaluate f_P directly but only P which is merely close and not equal to f_P. Self–testing takes care of the first stage while self–correcting is useful in the second stage. We shall illustrate the technique with our benchmark linearity testing problem. First, we state a result implicit in the proof of Theorem 3.

Corollary 2. *Let G and G' be two finite abelian groups, and $(e_i)_{1 \leq i \leq d}$ be a set of generators of G. Let $f : G \to G'$ be a homomorphism and $P : G \to G'$ be such that for some constant $\eta < 1/6$,*

$$\mathsf{Pr}_{x,y \in G}\left[P(x+y) - P(x) - P(y) \neq 0\right] \leq \eta.$$

Suppose also that $g(e_i) = f(e_i)$ for $i = 1, \ldots, d$, where by definition $g(x) = \mathsf{Maj}_{y \in G}\left(P(x+y) - P(y)\right)$. Then, $g = f$ and

$$\mathsf{Pr}_{x \in G}\left[f(x) \neq P(x)\right] \leq 2\eta.$$

Proof. Implicit in the proof of Theorem 3 is the fact that g is an homomorphism and that $\mathsf{Pr}_{x \in G}\left[g(x) \neq P(x)\right] \leq 2\eta$. Since two homomorphisms $f, g : G \to G'$ agree everywhere if and only if they agree on a set of generators of G, the desired conclusion follows.

The preceding result tells us that the following procedure leads to a continuous and robust exact test for the linear function f.

Specific Linear Function Test(P, f)
- **Linearity Test(P)**
  1. Randomly choose $x, y \in G$.
  2. Reject if $P(x+y) - P(x) - P(y) \neq 0$.
- **Generator Test(P)**
  1. For $i = 1, \ldots, d$, reject if $P(x + e_i) - P(x) - f(e_i) \neq 0$.

The second part of this procedure consists of verifying that the closest linear function to P coincides with f on a set of generators for the group domain. Thus, it illustrates an instance of the generator test. The generator test is done on the corrected program which is computed by the self–correcting process. Indeed, instead of comparing $f(e_i)$ and $P(e_i)$ the comparison is done with respect to $P(x + e_i) - P(x)$ for a randomly chosen x. This is necessary since P although close to an homomorphism $f' \neq f$ might agree with f over all generators — but, in this case $P(x + e_i) - P(x)$ will most likely agree with $f'(e_i)$ for a randomly chosen x. Finally, observe how the **Linearity Test** simplifies the task of verifying whether P is close to the homomorphism f. Indeed, when P is known to compute on a large fraction of the inputs an homomorphism g, it is sufficient to check that g equals f on a set of generators whose size can be very small (constant) compared to the size of the whole domain.

Corollary 3. *Using the notation of Corollary 2, the* **Specific Linear Function Test** *characterizes f, is $(\eta, (3+d)\eta)$–continuous, and $(6\eta, \eta)$–robust.*

Nonetheless, when the number of generators d is large (for example grows with the size of the group), the number of calls to the program in the **Generator Test** will be large. This situation is called the *generator bottleneck*.

2.5 Generator Bottleneck

In some cases, it is possible to get around the generator bottleneck using an *inductive test*. This is essentially another property test which eliminates the need of testing the self–corrected function on all the generators. We illustrate this point for the case of the Discrete Fourier Transform (DFT). The method and the results in this subsection are due to Ergün [Erg95]. For a more detailed discussion on the possibilities to circumvent the generator bottleneck by an inductive test see [ESK00].

Let p be a prime number and fix some $x = (x_0, \dots, x_n) \in \mathbb{Z}_p^{n+1}$, where $x_i \neq x_j$ for $i \neq j$. Then the linear function $\mathrm{DFT}_x : \mathbb{Z}_p^{n+1} \to \mathbb{Z}_p^{n+1}$ maps the coefficient representation $a = (a_0, \dots, a_n) \in \mathbb{Z}_p^{n+1}$ of a degree n polynomial $Q(X) = a_0 + a_1 X + \dots + a_n X^n$ into its point-value representation $(Q(x_0), \dots, Q(x_n)) \in \mathbb{Z}_p^{n+1}$. The group $\mathbb{Z}_p^{n+1}$ has $n + 1$ generators which can be chosen for example as $e_0 = (1, 0, \dots 0), \dots, e_n = (0, \dots, 0, 1)$. Applying the **Generator Test** in order to verify whether a linear function g is equal to DFT_x would require checking whether $g(e_i) = \mathrm{DFT}_x(e_i)$ for $i = 0, \dots, n$, and therefore the number of calls to the program would grow linearly with the degree of the polynomial. The key observation that helps to overcome this problem is that the generators $e_0, \dots, e_n$ can be obtained from each other by a simple linear operation and that the same is true for the values of DFT_x on $e_0, \dots, e_n$. We now explain in detail how to take advantage of this fact. First, we need to introduce some notation. For $a = (a_0, \dots, a_n) \in \mathbb{Z}_p^{n+1}$ let the rotation to the right vector be $\mathrm{ROR}(a) = (a_n, a_0, \dots, a_{n-1})$ and let $x \cdot a = (x_0 a_0, \dots, x_n a_n)$. Note that $e_{i+1} = \mathrm{ROR}(e_i)$ for $i = 0, \dots, n - 1$, that DFT_x maps e_0 to $(1, 1, \dots, 1)$, and most importantly, that DFT_x sends $\mathrm{ROR}(a)$ to $x \cdot \mathrm{DFT}_x(a)$ for all $a = (a_0, \dots, a_n) \in \mathbb{Z}_p^{n+1}$ with $a_n = 0$. Therefore, to verify whether a linear function g is equal to DFT_x it suffices to check whether g maps e_0 to $(1, 1, \dots, 1)$ and that $g(\mathrm{ROR}(a)) = x \cdot g(a)$ for all a with $a_n = 0$. The robustness of this testing procedure is guaranteed by the following:

Theorem 5. *Let $x \in \mathbb{Z}_p^{n+1}$ and let $P : \mathbb{Z}_p^{n+1} \to \mathbb{Z}_p^{n+1}$ be an application such that for some constant $\eta < 1/6$,*

$$\Pr_{a,b \in \mathbb{Z}_p^{n+1}} [P(a + b) - P(a) - P(b) \neq 0] \leq \eta,$$

$$\Pr_{c \in \mathbb{Z}_p^{n+1} : c_n = 0} [g(\mathrm{ROR}(c)) \neq x \cdot g(c)] < 1/2,$$

where $g(a) = \mathrm{Maj}_{b \in \mathbb{Z}_p^{n+1}} (P(a + b) - P(b))$ for all $a \in \mathbb{Z}_p^{n+1}$ and $g(1, 0, \dots, 0) = (1, \dots, 1)$. Then, $g = \mathrm{DFT}_x$ and

$$\Pr_{a \in \mathbb{Z}_p^{n+1}} [\mathrm{DFT}_x(a) \neq P(a)] \leq 2\eta.$$

Proof. Theorem 3 and the comment following its proof guarantee that g is linear and that P is close to g. The linearity of g implies that for every $a, b \in \mathbb{Z}_p^{n+1}$, we have $g(\mathrm{ROR}(a + b)) = g(\mathrm{ROR}(a)) + g(\mathrm{ROR}(b))$. By linearity we also have

$x \cdot (a + b) = x \cdot a + x \cdot b$ for every $a, b \in \mathbb{Z}_p^{n+1}$. Thus, the second probability bound in the hypotheses of the theorem implies that for all a with $a_n = 0$,

$$\Pr_{c \in \mathbb{Z}_p^{n+1}, c_n = 0} \left[g(\mathrm{ROR}(a)) = g(\mathrm{ROR}(c)) + g(\mathrm{ROR}(a - c)) = x \cdot g(a) \right] > 0.$$

Therefore, $g(\mathrm{ROR}(a)) = x \cdot g(a)$ always holds. To conclude the proof observe that the latter identity and the fact that $g(1, 0, \dots, 0) = (1, \dots, 1)$ imply that $g = \mathrm{DFT}_x$.

The previous result suggests the following exact test in order to ascertain whether a program computes DFT_x.

DFT_x Test(P)
1. Randomly choose $a, b \in \mathbb{Z}_p^{n+1}$.
2. Reject if $P(a + b) - P(a) - P(b) \neq 0$.
3. Reject if $P((1, 0, \dots, 0) + a) - P(a) - (1, \dots, 1) \neq 0$.
4. Randomly choose $c \in \mathbb{Z}_p^{n+1}$ such that $c_n = 0$.
5. Reject if $P(\mathrm{ROR}(c) + a) - P(a) - x \cdot (P(c + b) - P(b)) \neq 0$.

It follows that,

Corollary 4. *The* **DFT_x Test** *is such that it characterizes* DFT_x, *is* $(\eta, 6\eta)$–*continuous, and* $(6\eta, \eta)$–*robust.*

2.6 Beyond Self-Testing Linearity

So far we have discussed the testing problem for collections of linear functions. This was done for ease of exposition. The arguments and concepts we have described are also useful in testing non–linear functions. We now illustrate this fact with two examples.

Multiplication over $\mathbb{Z}_n$: The **Linearity Test** and the **Generator Test** can be combined to yield various self–testers. One such example allows to ascertain whether a program computes the multiplication function mult over $\mathbb{Z}_n$, i.e., the function that associates to $(x, y) \in \mathbb{Z}_n \times \mathbb{Z}_n$ the value xy (arithmetic over $\mathbb{Z}_n$). The exact test that achieves this goal is realized by the following procedure which is due to Blum, Luby, and Rubinfeld [BLR90]:

Multiplication Test(P)
1. Randomly choose $x, y, z \in \mathbb{Z}_n$.
2. Reject if $P(x, y + z) - P(x, y) - P(x, z) \neq 0$.
3. Reject if $P(x, y + 1) - P(x, y) - x \neq 0$.

Corollary 5. *The* **Multiplication Test** *is such that it characterizes* mult, *is* $(\eta, 4\eta)$–*continuous, and* $(6\eta, \eta)$–*robust.*

Proof. For a fixed $x \in \mathbb{Z}_n$, let $l_x : \mathbb{Z}_n \to \mathbb{Z}_n$ be the linear function defined by $l_x(y) = xy$. Let d_x be the distance $\mathsf{Dist}(P(x, \cdot), l_x)$ and let e_x be the rejection probability of the test for randomly chosen $y, z \in \mathbb{Z}_n$. By Corollary 3, we know that $e_x/4 \le d_x \le 6e_x$ for all x. Observe now that $\mathsf{E}_{x \in \mathbb{Z}_n}[d_x] = \mathsf{Dist}(P, \mathsf{mult})$ and that $\mathsf{E}_{x \in \mathbb{Z}_n}[e_x]$ is the probability that the test rejects P. The desired result follows.

Polynomials: Let $\mathbb{F}$ be a field and let $f : \mathbb{F} \to \mathbb{F}$ be a function. We adopt the standard convention of denoting the *forward difference operator* by ∇_t. Hence, by definition, $\nabla_t f(x) = f(x + t) - f(x)$ for $x, t \in \mathbb{F}$. If we let ∇_t^d denote the operator corresponding to d applications of ∇_t and for $\boldsymbol{t} \in \mathbb{F}^d$ denote by $\nabla_{\boldsymbol{t}}$ the operator corresponding to the applications of $\nabla_{t_1}, \ldots, \nabla_{t_d}$, then it is easy to check that:

1. ∇_t is linear,
2. ∇_{t_1} and ∇_{t_2} commute,
3. $\nabla_{t_1, t_2} = \nabla_{t_1 + t_2} - \nabla_{t_1} - \nabla_{t_2}$, and
4. $\nabla_t^d = \sum_{k=0}^{d} (-1)^{d-k} \binom{d}{k} \nabla_{kt}$.

The usefulness of the difference operator in testing was recognized by Rubinfeld and Sudan [RS92b]. They used it to give a more efficient self–corrector for polynomials over finite fields than the one proposed by Lipton [Lip91]. Its utility is mostly based on two facts: $\nabla_t f(x)$ can be computed efficiently, and it gives rise to the following well known characterization of polynomials:

Theorem 6. *Let p be a prime number, let $f : \mathbb{Z}_p \to \mathbb{Z}_p$ be a function, and let $d < p - 1$. Then f is a degree d polynomial over $\mathbb{Z}_p$ if and only if $\nabla_t^{d+1} f(x) = 0$ for all $x, t \in \mathbb{Z}_p$.*

The preceding theorem gives a functional equation characterization of degree d polynomials. Hence, it gives rise to the following functional equation test:

Degree d Polynomial Test(P)
1. Randomly pick $x, t \in \mathbb{Z}_p$.
2. Reject if $\nabla_t^{d+1} P(x) \ne 0$.

The above described exact test was proposed and analyzed in [RS92b]. Let us now discuss shortly its properties. For the sake of simplicity, consider the following particular case where $d = 1$:

Affine Test(P)
1. Randomly pick $x, t \in \mathbb{Z}_p$.
2. Reject if $P(x + 2t) - 2P(x + t) + P(x) \ne 0$.

Instead of choosing t as above it is tempting to pick two values t_1 and t_2 also in $\mathbb{Z}_p$ and check whether $P(x + t_1 + t_2) - P(x + t_1) - P(x + t_2) + P(x) \ne 0$. This is not an acceptable verification procedure in the self–testing context since it is essentially equivalent to affine interpolation (polynomial interpolation in the general case). Hence, it is not really computationally simpler than computing the functions of the class one wishes to test. On the contrary, the **Degree** d **Polynomial Test** is computationally more efficient than computing a degree d polynomial. Moreover, it requires less evaluations of the program P. This justifies the use of the ∇_t^{d+1} operator in testing degree d polynomials.

Since the **Degree** d **Polynomial Test** is $(d+2)$–local, the standard approach for proving continuity of such tests yield that it is $(\eta, (d+2)\eta)$–continuous. The robustness of the test is guaranteed by the following result of Rubinfeld and Sudan [RS92b]:

Theorem 7. *Let p be a prime number, let $P : \mathbb{Z}_p \to \mathbb{Z}_p$ be a function, and let $d < p - 1$. If for some $\eta < 1/2(d+2)^2$,*

$$\Pr_{x,t \in \mathbb{Z}_p} \left[\nabla_t^{d+1} P(x) \neq 0 \right] \leq \eta,$$

then there exists a degree d polynomial $g : \mathbb{Z}_p \to \mathbb{Z}_p$ such that

$$\Pr_{x \in \mathbb{Z}_p} \left[g(x) \neq P(x) \right] \leq 2\eta.$$

Proof (Sketch). The proof is a standard application of the majority function argument albeit algebraically somewhat involved. We only describe the main proof steps. For $i = 0, \ldots, d+1$, let $\alpha_i = (-1)^{i+1} \binom{d+1}{i}$. Note that $\nabla_t^{d+1} P(x) = 0$ if and only if $P(x) = \sum_{i=1}^{d+1} \alpha_i P(x+it)$. This motivates the following definition:

$$g(x) = \mathsf{Maj}_{t \in \mathbb{Z}_p} \left(\sum_{i=1}^{d+1} \alpha_i P(x+it) \right).$$

The proof proceeds in the typical three stages that application of the majority function argument gives rise to. First, one shows that with overwhelming probability $\sum_{i=1}^{d+1} \alpha_i P(x+it)$ agrees with $g(x)$, in particular that

$$\Pr_{t \in \mathbb{Z}_p} \left[g(x) = \sum_{i=1}^{d+1} \alpha_i P(x+it) \right] \geq 1 - 2(d+1)\eta.$$

Second, one establishes that the distance between g and P is at most 2η. Finally, one proves that

$$\Pr_{t_1,t_2 \in \mathbb{Z}_p} \left[\sum_{i=0}^{d+1} \alpha_i g(x+it) = 0 \right] \geq 1 - 2(d+2)^2 \eta > 0$$

Since the event $\sum_{i=0}^{d+1} \alpha_i g(x+it) = 0$ is independent of t_1 and t_2, we get that $\sum_{i=0}^{d+1} \alpha_i g(x+it) = 0$ must hold. The desired conclusion follows.

3   Approximate Self-Testing

Initially it was assumed in the self–testing literature that programs performed exact computations and that the space of valid inputs was closed under the standard arithmetic operations, i.e., was an algebraically closed domain. However, early on it was recognized that these assumptions were too simplistic to capture the real nature of many computations, in particular the computation of real valued functions and of functions defined over finite rational domains (finite subsets of fixed point arithmetic of the form $\{i/s \ : \ |i| \leq n, i \in \mathbb{Z}\}$ for some $n, s > 0$).

Self–testers/correctors for programs whose input values are from finite rational domains were first considered by Lipton [Lip91] and further developed by Rubinfeld and Sudan [RS92b]. In [Lip91] a self–corrector for multivariate polynomials over a finite rational domain is given. In the same scenario [RS92b] describes more efficient versions of this result as well as a self–tester for univariate polynomials. The study of self–testing in the context of inexact computations was started by Gemmell et al. [GLR+91] who provided approximate self–testers/correctors for linear functions, logarithmic functions, and floating point exponentiation. Nevertheless, their work was limited to the context of algebraically closed domains. Program checking in the approximate setting was first considered by Ar et al. [ABCG93] who provided, among others, approximate checkers for some trigonometric functions and matrix operations. Considering both aspects simultaneously led to the development of approximate self–testers over finite rational domains by Ergün, Kumar, and Rubinfeld [EKR96]. Among other things, they showed how to perform approximate self–testing with absolute error for linear functions, polynomials, and for functions satisfying addition theorems.

We now begin a formal discussion of the theory of approximate testing. Our exposition follows the presentation in [Mag00a] which has the advantage of encompassing all models of approximations studied in the self–testing literature.

Throughout this section, let D be a finite set and let R be a metric space. The distance in R will be denoted by $d(\cdot, \cdot)$, When R is also a normed space, its norm will be denoted by $|| \cdot ||$. As usual we denote by $\mathcal{C}$ the family of functions from D to R.

As in the case of the exact testing problem we are again interested in determining, maybe probabilistically, how "close" a program $P : D \to R$ is to an underlying family of functions of interest $\mathcal{F} \subseteq \mathcal{C}$. But now, the elements of R might be hard to represent (for example, when $\mathcal{F}$ is a family of trigonometric functions). Thus, any reasonable program P for computing $f \in \mathcal{F}$ will necessarily have to compute an approximation. In fact, P might never equal f over D but still be, for all practical purposes, a good computational realization of a program that computes f. Hence, the way in which we captured the notion of "closeness" in Section 2, that is, Definition 1, is now inadequate. Thus, to address the testing problem for R valued functions we need a different notion of incorrect computation. In fact, we need a definition of error. This leads to the following:

Definition 6 (Computational error term). *A computational error term for C is a function $\varepsilon : D \times R \to \mathbb{R}^+$. If $P, f : D \to R$ are two functions, then P ε-computes f on $x \in D$ if $d(P(x), f(x)) \leq \varepsilon(x, f(x))$.*

This definition encompasses several models of approximate computing that depend on the restriction placed on the computational error term ε. Indeed, it encompasses the

- exact computation case, where $\varepsilon(x, v) = 0$,
- approximate computation with absolute error, where $\varepsilon(x, v) = \varepsilon_0$ for some constant $\varepsilon_0 \in \mathbb{R}^+$,
- approximate computation with error relative to input size, where $\varepsilon(x, v) = \varepsilon_1(x)$ for some function $\varepsilon_1 : D \to \mathbb{R}^+$ depending only on x,
- approximate computation with relative error, where R is a normed space and $\varepsilon(x, v) = \theta||v||$ for some constant $\theta \in \mathbb{R}^+$.

Based on the definition of computational error term we can give a notion of distance, similar to that of Definition 1, which is more appropriate for the context of approximate computation.

Definition 7 (ε-Distance). *Let $P, f \in C$, let $D' \subseteq D$, and let ε be a computational error term. The ε-distance of P from f on D' is[2]*

$$\mathrm{Dist}(P, f, D', \varepsilon) = \mathrm{Pr}_{x \in D'} \left[P \text{ does not } \varepsilon\text{-compute } f \text{ on } x \right].$$

If $\mathcal{F} \subseteq C$, then the ε-distance of P from $\mathcal{F}$ on D' is

$$\mathrm{Dist}(P, \mathcal{F}, D', \varepsilon) = \inf_{f \in \mathcal{F}} \mathrm{Dist}(P, f, D', \varepsilon).$$

The new notion of distance naturally gives rise to extensions of the notions introduced in Section 2. In what follows, we state these extensions.

Definition 8 (Approximate self-tester). *Let $\mathcal{F} \subseteq C$, let $D' \subseteq D$, let ε and ε' be computational error terms and let $0 \leq \eta \leq \eta' < 1$ be constants. A $(D, \varepsilon, \eta; D', \varepsilon', \eta')$-(approximate) self-tester for $\mathcal{F}$ on C is a probabilistic oracle Turing machine T such that for every $P \in C$ and for every confidence parameter $0 < \gamma < 1$:*

- *if $\mathrm{Dist}(P, \mathcal{F}, D, \varepsilon) \leq \eta$, then $\mathrm{Pr}\left[T^P(\gamma) = \texttt{GOOD} \right] \geq 1 - \gamma$;*
- *if $\mathrm{Dist}(P, \mathcal{F}, D', \varepsilon') > \eta'$, then $\mathrm{Pr}\left[T^P(\gamma) = \texttt{BAD} \right] \geq 1 - \gamma$,*

where the probabilities are taken over the coin tosses of T.

Definition 9 (Approximate test). *An approximate test $(\mathcal{A}, \mathcal{C}, \mathcal{D}, \beta)$ is a set of applications $\mathcal{A}$ from C to $\mathbb{R}^+$ with a distribution $\mathcal{D}$ over $\mathcal{A}$ and a test error,*

---

[2] The need for considering the values taken by P over a subset D' of f's domain is a technical one. We discuss this issue later on. In the meantime, the reader might prefer to simply assume $D' = D$.

i.e., a function $\beta : \mathcal{A} \times \mathcal{C} \to \mathbb{R}^+$. *The approximate test* characterizes *the family of functions*

$$\mathsf{char}(\mathcal{A}, \mathcal{C}, \mathcal{D}) = \{f \in \mathcal{C} \ : \ \mathsf{Pr}_{t \in_{\mathcal{D}} \mathcal{A}}\left[t(f) = 0\right] = 1\}.$$

The rejection probability *of a function* $P \in \mathcal{C}$ *by the approximate test is defined as*

$$\mathsf{Rej}(P, \mathcal{A}, \beta) = \mathsf{Pr}_{t \in_{\mathcal{D}} \mathcal{A}}\left[t(P) > \beta(t, P)\right].$$

A probabilistic oracle Turing machine M *realizes the approximate test if for all* $P \in \mathcal{C}$,

$$\mathsf{Pr}\left[M^P \ \text{returns BAD}\right] = \mathsf{Rej}(P, \mathcal{A}, \beta),$$

where the probability on the left hand side is taken over the internal coin tosses of the machine M.

As in the case of exact testing, we specify approximate tests through the following high level description:

Approximate Test$(P \in \mathcal{C}, \mathcal{A}, \mathcal{D}, \beta)$
1. Choose an element $t \in \mathcal{A}$ according to $\mathcal{D}$.
2. Reject if $t(P) > \beta(t, P)$.

Note that one needs to compute the test error for realizing the approximate test. Also, exact tests are a particular case of approximate tests where the test error is 0 everywhere, GOOD is identified with 0 and BAD with 1.

In order not to unnecessarily clutter the notation, we again omit $\mathcal{C}$, $\mathcal{D}$, and β whenever clear from context and restrict our discussion to the case where $\mathcal{D}$ is the uniform distribution.

The robustness and continuity properties of exact tests are now generalized as follows:

Definition 10 (Continuity & robustness). *Let* ε *be a computational error term for* $\mathcal{C}$, *let* $D' \subseteq D$, *and let* $(\mathcal{A}, \beta)$ *be an approximate test on* $\mathcal{C}$ *which characterizes the family* $\mathcal{F}$. *Also, let* $0 \leq \eta, \delta < 1$ *be constants. Then,* $(\mathcal{A}, \beta)$ *is* (η, δ)–continuous *on* D' *with respect to* ε *if for all* $P \in \mathcal{C}$,

$$\mathsf{Dist}(P, \mathcal{F}, D', \varepsilon) \leq \eta \Longrightarrow \mathsf{Rej}(P, \mathcal{A}, \beta) \leq \delta,$$

and it is (η, δ)–robust *on* D' *with respect to* ε *if for all* $P \in \mathcal{C}$,

$$\mathsf{Rej}(P, \mathcal{A}, \beta) \leq \delta \Longrightarrow \mathsf{Dist}(P, \mathcal{F}, D', \varepsilon) \leq \eta.$$

The continuity and the robustness of an approximate test give rise to the construction of approximate self–testers through the following:

Theorem 8 (Approximate generic self–tester). *Let* $0 \leq \delta < \delta' < 1$ *and* $0 \leq \eta \leq \eta' < 1$ *be constants,* $\mathcal{C}$ *be a family of functions from a finite set* D *to a metric space* R, ε *and* ε' *be computational error terms for* $\mathcal{C}$, *and* $D' \subseteq D$. *Also, let* $(\mathcal{A}, \beta)$ *be an approximate test on* $\mathcal{C}$ *that is realized by the probabilistic Turing machine* M. *If* $(\mathcal{A}, \beta)$ *characterizes the family* $\mathcal{F}$,

- *is (η, δ)–continuous on D with respect to ε, and*
- *(η', δ')–robust on D' with respect to ε',*

then there exists a $(D, \varepsilon, \eta; D', \varepsilon', \eta')$–self-tester for $\mathcal{F}$ on $\mathcal{C}$ which performs for every confidence parameter $0 < \gamma < 1$, $O(\ln(1/\gamma)\frac{\delta + \delta'}{(\delta - \delta')^2})$ iterations of M, counter increments, comparisons, and binary shifts.

Proof. Similar to the proof of Theorem 1.

As in the case of exact self–testing, realizable approximate tests are often constructed through functional equations. Specifically, for $D' \subseteq D$, let $\Phi : \mathcal{C} \times \mathcal{N} \to \mathbb{R}$ be a functional where $\mathcal{N} \subseteq \bigcup_{k=1}^{|D'|}(D')^k$ is a collection of neighborhoods. The functional Φ and a function $\beta' : \mathcal{N} \to \mathbb{R}^+$ induce an approximate test $(\mathcal{A}, \beta)$ by defining for all $(x_1, \dots, x_k) \in \mathcal{N}$ the mapping $t_{x_1, \dots, x_k}(f) : \mathcal{F} \to \mathbb{R}^+$ as $t_{x_1, \dots, x_k}(f) = |\Phi(f, x_1, \dots, x_k)|$, making $\beta(t_{x_1, \dots, x_k}, f) = \beta'(x_1, \dots x_k)$, and letting

$$\mathcal{A} = \{t_{x_1, \dots, x_k} : (x_1, \dots, x_k) \in \mathcal{N}\}.$$

By definition,

$$\mathrm{char}(\mathcal{A}) = \{f \in \mathcal{C} : \forall (x_1, \dots, x_k) \in \mathcal{N}, \quad \Phi(P, x_1, \dots, x_k) = 0\}.$$

If Φ and β' are efficiently computable, then a Turing machine M realizes the induced approximate test by choosing $(x_1, \dots x_k) \in \mathcal{N}$ and comparing the value $|\Phi(f, x_1, \dots, x_k)|$ to $\beta'(x_1, \dots x_k)$. When $(\mathcal{A}, \beta)$ is continuous and robust with respect to some computational error term, Theorem 8 can be applied to derive a corresponding approximate self–tester. The complexity of the self–tester will ultimately depend on the complexity of computing Φ and β'.

The approximate testing problem is technically more challenging and involved than the exact testing problem. We shall try to smoothly introduce the new aspects that one encounters in the approximate testing scenario. Thus, the discussion that follows is divided into three parts: the case of absolute error, the case of error relative to the input size, and finally the case of relative error. The discussion becomes progressively more involved. We shall try to stress the common arguments used in the different cases, but will discuss each one in a separate section.

Before proceeding, it is worth pointing out two common aspects of all known analyses of approximate tests. Specifically, in their proofs of robustness. First, they are considerably more involved than in the case of exact testing. Second, there are two clearly identifiable stages in such proofs. In each stage, it is shown that the approximate test exhibits one of the properties captured by the following two notions:

Definition 11 (Approximate robustness). *Let ε be a computational error term for $\mathcal{C}$ and $D' \subseteq D$. Let $(\mathcal{A}, \beta)$ and $(\mathcal{A}', \beta')$ be approximate tests on $\mathcal{C}$, both characterizing the family $\mathcal{F}$. Let $0 \le \eta, \delta < 1$ be constants. Then, $(\mathcal{A}, \beta)$ is (η, δ)–approximately robust for $(\mathcal{A}', \beta')$ on D' with respect to ε if for all $P \in \mathcal{C}$,*

$$\mathsf{Rej}(P, \mathcal{A}, \beta) \le \delta \implies \exists g \in \mathcal{C}, \ \mathsf{Dist}(P, g, D', \varepsilon) \le \eta, \ \mathsf{Rej}(g, \mathcal{A}', \beta') = 0.$$

Definition 12 (Stability). *Let ε be a computational error term for $\mathcal{C}$ and $D' \subseteq D$. Let $(\mathcal{A}, \beta)$ be an approximate test on $\mathcal{C}$ which characterizes the family $\mathcal{F}$. Then, $\mathcal{A}$ is* stable *on D' with respect to ε if for all $g \in \mathcal{C}$,*

$$\mathsf{Rej}(g, \mathcal{A}, \beta) = 0 \Longrightarrow \mathsf{Dist}(g, \mathcal{F}, D', \varepsilon) = 0.$$

Note that stability is nothing else than $(0,0)$–robustness. A direct consequence of these definitions is that if $(\mathcal{A}, \beta)$ is approximately robust for $(\mathcal{A}', \beta')$ with respect to ε and $(\mathcal{A}', \beta')$ is stable with respect to ε', then $(\mathcal{A}, \beta)$ is also robust with respect to $\varepsilon + \varepsilon'$.

We henceforth restrict our discussion to real valued functions whose domain is $D_n = \{i \in \mathbb{Z} : |i| \leq n\}$ for some $n > 0$. Our results can be directly extended to finite rational domains. We conclude this section by stating some general facts that play a key role in the design and analysis of all approximate tests.

3.1 Basic Tools

Here we state two simple lemmas which will be repeatedly applied in the forthcoming sections.

Definition 13 (Median). *For $f : X \to \mathbb{R}$ denote by $\mathsf{Med}_{x \in X} (f(x))$ the median of the values taken by f when x varies in X, i.e.,*

$$\mathsf{Med}_{x \in X} (f(x)) = \mathsf{Inf}\{a \in \mathbb{R} : \mathsf{Pr}_{x \in X} [f(x) > a] \leq 1/2\}.$$

Lemma 1 (Median principle). *Let D, D' be two finite sets. Let $\varepsilon \geq 0$ and $F : D \times D' \to \mathbb{R}$. Then,*

$$\mathsf{Pr}_{x \in D} [|\mathsf{Med}_{y \in D'} (F(x, y))| > \varepsilon] \leq 2\mathsf{Pr}_{(x,y) \in D \times D'} [|F(x, y)| > \varepsilon].$$

Proof. Observe that

$$\mathsf{Pr}_{x \in D} [|\mathsf{Med}_{y \in D'} (F(x, y))| > \varepsilon] \leq \mathsf{Pr}_{x \in D} [\mathsf{Pr}_{y \in D'} [|F(x, y)| > \varepsilon] > 1/2],$$

and apply Markov's inequality.

Lemma 2 (Halving principle). *Let Ω and S denote finite sets such that $S \subseteq \Omega$, and let ψ be a boolean function defined over Ω. Then,*

$$\mathsf{Pr}_{x \in S} [\psi(x)] \leq \frac{|\Omega|}{|S|} \mathsf{Pr}_{x \in \Omega} [\psi(x)].$$

Proof.

$$\mathsf{Pr}_{x \in \Omega} [\psi(x)] \geq \mathsf{Pr}_{x \in \Omega} [\psi(x)|x \in S] \, \mathsf{Pr}_{x \in \Omega} [x \in S] = \frac{|S|}{|\Omega|} \mathsf{Pr}_{x \in \Omega} [\psi(x)].$$

If Ω is twice the size of S, then $\mathsf{Pr}_{x \in \Omega} [\psi(x)]$ is at least one half of $\mathsf{Pr}_{x \in S} [\psi(x)]$. This motivates the choice of name for Lemma 2.

We will soon see the importance that the median function has in the context of approximate self–testing. This was recognized by Ergün, Kumar, and Rubinfeld in [EKR96] where the median principle was also introduced. The fact that the Halving principle can substantially simplify the standard proof arguments one encounters in the approximate testing scenario was observed in [KMS99].

4   Testing with Absolute Error

Throughout this section we follow the notation introduced in the previous one. Moreover, we restrict our discussion to the case of absolute error, i.e., to the case where $\varepsilon(x, v)$ is some non–negative real constant ε. Again, for the purpose of illustration we consider the linearity testing problem over a rational domain D, say $D = D_{8n}$ for concreteness. Hence, taking $D' = D_{4n}$, the functional equation

$$\forall x, y \in D_{4n}, \qquad P(x + y) - P(x) - P(y) = 0,$$

gives rise to the following approximate absolute error test:

Absolute error Linearity Test(P, ε)
1. Randomly choose $x, y \in D_{4n}$.
2. Reject if $|P(x + y) - P(x) - P(y)| > \varepsilon$.

The preceding approximate test was proposed and analyzed by Ergün, Kumar, and Rubinfeld [EKR96]. We illustrate the crucial issues related to testing under absolute error by fully analyzing this approximate test. Our discussion is based on [EKR96] and simplifications proposed in [KMS99].

4.1   Continuity

As in the case of exact testing, continuity is a property which is usually much easier to establish than robustness. Although proofs of continuity in the approximate case follow the same argument than in the exact case, there is a subtlety involved. It concerns the use of the Halving principle as shown by the following result from which $(\eta, 6\eta)$–continuity of the **Absolute error Linearity Test** immediately follows.

Lemma 3. *Let $\varepsilon \geq 0$. Let P, l be real valued functions over D_{8n} such that l is linear. Then,*

$$\Pr_{x,y \in D_{4n}} \left[|P(x + y) - P(x) - P(y)| > 3\varepsilon\right] \leq 6\Pr_{x \in D_{8n}} \left[|P(x) - l(x)| > \varepsilon\right].$$

Proof. Simply observe that $|P(x+y) - P(x) - P(y)| > 3\varepsilon$ implies $|P(x+y) - l(x+y)| > \varepsilon$ or $|P(x) - l(x)| > \varepsilon$ or $|P(y) - l(y)| > \varepsilon$. By the Halving principle, the probability that each of these three events occur when x and y are independently and uniformly chosen in D_{4n} is at most $2\Pr_{x \in D_{8n}} \left[|P(x) - l(x)| > \varepsilon\right]$. Thus, the union bound yields the desired conclusion.

4.2   Approximate Robustness

We now describe how robustness is typically established. Our discussion is based on [EKR96]. The majority argument will again be useful, but it needs to be modified. To see why, recall that the argument begins by defining a function g

whose value at x takes the most commonly occurring value among the members of a multiset S_x whose elements depend on x and P, i.e.,

$$g(x) = \mathsf{Maj}_{s \in S_x} (s) \, .$$

Each value in S_x is seen as an estimation of the correct value of P on x. But now, P is not restricted to taking a finite number of values. There might not be any clear majority, or even worse, all but one pair of values in every set S_x might be distinct while very different from all other values in the set — the latter of these values might even be very similar among themselves. Thus, the $\mathsf{Maj}\,(\cdot)$ is not a good estimator in the context of testing programs that only approximately compute the desired value. A more robust estimator is needed. This explains why $\mathsf{Med}\,(\cdot)$ is used instead of $\mathsf{Maj}\,(\cdot)$. This gives rise to what we shall call the median function argument. The robustness proofs based on it will also exhibit three stages. The first two are similar to those encountered in the majority function argument. Indeed, first one shows that an overwhelming number of the elements of S_x are good approximations of $g(x) = \mathsf{Med}_{s \in S_x} (s)$, then one shows that g is close to P. The major difference is in the third stage — it falls short of establishing that g has the property one is interested in. For the sake of concreteness, we now illustrate what happens in the case of linearity testing.

Theorem 9. *Let $\varepsilon \geq 0$ and $0 \leq \eta < 1/96$ be constants and let $P : D_{8n} \to \mathbb{R}$ be an application such that*

$$\mathsf{Pr}_{x,y \in D_{4n}} \left[|P(x+y) - P(x) - P(y)| > \varepsilon \right] \leq \eta.$$

Then, there exists a function $g : D_{2n} \to \mathbb{R}$ such that

$$\mathsf{Pr}_{x \in D_n} \left[|g(x) - P(x)| > \varepsilon \right] \leq 16\eta,$$

and for all $a, b \in D_n$,

$$|g(a+b) - g(a) - g(b)| \leq 6\varepsilon.$$

Proof. Let $P_{x,y} = P(x+y) - P(x) - P(y)$. Define the function $g : D_{2n} \to \mathbb{R}$ by $g(x) = \mathsf{Med}_{y \in D_{2n}} (P(x+y) - P(y))$. First, we show that with overwhelming probability $P(x+y) - P(y)$ is a good approximation to $g(x)$, specifically, that for all $c \in D_{2n}$ and $I \subseteq D_{2n}$ such that $|I| = |D_n|$,

$$\mathsf{Pr}_{y \in I} \left[|g(c) - (P(c+y) - P(y))| > 2\varepsilon \right] < 32\eta. \tag{2}$$

The Median principle implies that

$$\mathsf{Pr}_{y \in I} \left[|g(c) - (P(c+y) - P(y))| > 2\varepsilon \right] \leq 2\mathsf{Pr}_{y \in I, z \in D_{2n}} \left[|P_{c+y,z} - P_{c+z,y}| > 2\varepsilon \right].$$

Observe that if y and z are randomly chosen in I and D_{2n} respectively, then the union bound yields

$$\mathsf{Pr}_{y,z} \left[|P_{c+y,z} - P_{c+z,y}| > 2\varepsilon \right] \leq \mathsf{Pr}_{y,z} \left[|P_{c+z,y}| > \varepsilon \right] + \mathsf{Pr}_{y,z} \left[|P_{c+y,z}| > \varepsilon \right].$$

To obtain (2), note that the Halving principle implies that the latter sum is at most

$$2\frac{|D_{4n}|^2}{|D_n||D_{2n}|}\mathsf{Pr}_{x,y\in D_{4n}}\left[|P_{x,y}|>\varepsilon\right].$$

To see that g is close to P, observe that the Halving principle implies that

$$\mathsf{Pr}_{x\in D_n}\left[|g(x)-P(x)|>\varepsilon\right]\leq 4\mathsf{Pr}_{x\in D_{4n}}\left[|g(x)-P(x)|>\varepsilon\right].$$

By definition of g we get that $g(x)-P(x)=\mathsf{Med}_{y\in D_{2n}}\left(P_{x,y}\right)$. Hence, the Median principle and the Halving principle yield

$$\mathsf{Pr}_{x\in D_n}\left[|g(x)-P(x)|>\varepsilon\right]\leq 8\mathsf{Pr}_{x\in D_{4n},y\in D_{2n}}\left[|P_{x,y}|>\varepsilon\right]$$
$$\leq 8\frac{|D_{4n}|}{|D_{2n}|}\mathsf{Pr}_{x,y\in D_{4n}}\left[|P_{x,y}|>\varepsilon\right].$$

Elementary calculations and the hypothesis imply that the last expression is upper bounded by 16η.

Finally, let $a,b\in D_n$. Three applications of (2) imply that for some $y\in D_n$

$$|g(a)-(P(a+y)-P(y))|\leq 2\varepsilon,$$
$$|g(b)-(P(a+b+y)-P(a+y))|\leq 2\varepsilon,$$
$$|g(a+b)-(P(a+b+y)-P(y))|\leq 2\varepsilon.$$

It follows that $|g(a+b)-g(a)-g(b)|\leq 6\varepsilon$.

The previous result falls short of what one desires. Indeed, it does not show that a low rejection probability for the **Absolute error Linearity Test** guarantees closeness to linearity. Instead, it establishes that if

$$|P(x+y)-P(x)-P(y)|>\varepsilon$$

holds for most x's and y's in a large domain, then P must be close to a function g which is approximately linear, i.e., for all a's and b's in a small domain,

$$|g(a+b)-g(a)-g(b)|\leq 6\varepsilon.$$

A conclusion stating that $g(a+b)=g(a)+g(b)$ would have been preferable. This will follow by showing that g is close to a linear function, thus implying the closeness of P to a linear function. By Definition 12, these results whereby it is shown that a function that approximately satisfies a functional equation everywhere must be close to a function that exactly satisfies the functional equation, are called stability proofs. Also, by Definition 11, results as those we have shown so far (i.e., whereby it is proved that a function that approximately satisfies a functional equation for most inputs must be close to a function that approximately satisfies the functional equation everywhere) are called approximate robustness proofs. As mentioned earlier, approximate robustness and stability imply robustness. In the following section we discuss a technique for proving stability results.

4.3 Stability

The main result of this section, i.e., the statement concerning stability of the **Absolute error Linearity Test** is from [EKR96]. However, the proof presented here is from [KMS99] and is based on an argument due to Skopf [Sko83]. The proof technique is also useful for obtaining stability results in the context of approximate testing over finite rational domains. It relies on two ideas developed in the context of stability theory. The first consists in associating to a function g approximately satisfying a functional equation a function h approximately satisfying the same functional equation but over an algebraically closed domain, e.g., a group. The function h is carefully chosen so that h agrees with g over a given subset of g's domain. In other words, h will be an *extension* of g. Thus, showing that h can be well approximated by a function with a given property is sufficient to establish that the function g can also be well approximated by a function with the same property. This task is easier to address due to the fact that h's domain has a richer algebraic structure. In fact, there is a whole community that for over half a century has been dedicated to the study of these type of problems. Indeed, in 1941, Hyers [Hye41] addressed one such problem for functions whose domain have a semi–group structure. The work of Hyers was motivated by a question posed by Ulam. Coincidentally, Ulam's question concerned linear functions. Specifically, Ulam asked whether a function f that satisfies the functional equation $f(x + y) = f(x) + f(y)$ only approximately could always be approximated by a linear function. Hyers showed that f could be approximated within a constant error term by a linear function when the equality was correct also within a constant term. To be precise, Hyers proved the following:

Theorem 10 (Hyers). *Let E_1 be a normed semi–group, let E_2 be a Banach space, and let $h : E_1 \to E_2$ be a mapping such that for all $x, y \in E_1$,*

$$\|h(x + y) - h(x) - h(y)\| \le \varepsilon.$$

Then, the function $l : E_1 \to E_2$ defined by $l(x) = \lim_{m \to \infty} h(2^m x)/2^m$ is a well defined linear mapping such that for all $x \in E_1$,

$$\|h(x) - T(x)\| \le 2\varepsilon.$$

Remark 1. We have stated Theorem 10 in its full generality in order to highlight the properties required of the domain and range of the functions we deal with. Also for this purpose, as long as we discuss stability issues, we keep the exposition at this level of generality. Nevertheless, we will apply Theorem 10 only in cases where $E_1 = \mathbb{Z}$ and $E_2 = \mathbb{R}$.

Many other Ulam type questions have been posed and satisfactorily answered. For surveys of such results see [HR92,For95]. But, these results cannot directly be applied in the context of approximate testing. To explain this, recall that we are concerned with functions g such that $|g(x + y) - g(x) - g(y)| \le \varepsilon$ only for x's and y's in D_n — which is not a semi–group. To address this issue and exploit results like those of Hyers one associates to g a function h that extends it over a larger domain which is typically a group. Moreover, the extension is done in such a way that one can apply a Hyers's type theorem.

Although the approach described in the previous paragraph is a rather natural one, it requires more work than necessary, at least for our purposes. Indeed, when deriving a stability type result for the approximate testing problem over D_n one considers the extension h of g given by $h(x) = g(r_x) + q_x g(n)$, where $q_x \in \mathbb{Z}$ and $r_x \in D_n$ are the unique numbers such that $x = q_x n + r_x$ and $|q_x n| < |x|$ if $x \in \mathbb{Z} \setminus \{0\}$, and $q_0 = r_0 = 0$. (See Fig. 2.) Thus, the limit of $h(2^m x)/2^m$ when m goes to ∞ is $x g(n)/n$. Hence, there is no need to prove that $l(x) = \lim_{m \to \infty} h(2^m x)/2^m$ is well defined and determines a linear mapping. Thus, when Hyers's theorem is applied to a function like h the only new thing we get is that l is close to h. As shown in the next lemma, to obtain this same conclusion a weaker hypothesis than that of Theorem 10 suffices. This fact significantly simplifies the proof of the stability results needed in the context of approximate testing.

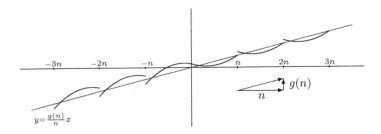

Fig. 2. Extension of g.

Lemma 4. *Let E_1 be a normed semi–group and E_2 be a Banach space. Let $\varepsilon \geq 0$ and let $h : E_1 \to E_2$ be such that for all $x \in E_1$,*

$$\|h(2x) - 2h(x)\| \leq \varepsilon.$$

Then, the function $T : E_1 \to E_2$ defined by $T(x) = \lim_{m \to \infty} h(2^m x)/2^m$ is a well defined mapping such that for all $x \in E_1$,

$$\|h(x) - T(x)\| \leq \varepsilon.$$

Proof. We follow the argument used by Hyers [Hye41] to prove Lemma 4. First, we show by induction on m, that

$$\left\| \frac{h(2^m x)}{2^m} - h(x) \right\| \leq \varepsilon \sum_{t=1}^{m} 2^{-t}. \tag{3}$$

The case $m = 1$ holds due to the hypothesis. Assume the claim is true for m. To prove the claim for $(m + 1)$, note that

$$\left\| \frac{h(2^{m+1}x)}{2^{m+1}} - h(x) \right\| \leq \left\| \frac{h(2x)}{2} - h(x) \right\| + \frac{1}{2} \left\| \frac{h(2^m \cdot 2x)}{2^m} - h(2x) \right\|$$

$$\leq \frac{\varepsilon}{2} + \frac{\varepsilon}{2} \sum_{t=1}^{m} 2^{-t}$$

$$= \varepsilon \sum_{t=1}^{m+1} 2^{-t}.$$

Fix $x = 2^k y$ in (3). Then, the sequence $(h(2^k y)/2^k)_k$ satisfies a Cauchy criterion for every y. Therefore, T is well defined. Letting $m \to \infty$ in (3) one obtains the desired conclusion.

Thus, to establish the stability type result we are seeking for the linearity testing problem one needs to show that an appropriate extension $h : \mathbb{Z} \to \mathbb{R}$ of a function $g : D_{2n} \to \mathbb{R}$ such that $|g(x+y) - g(x) - g(y)| \leq \varepsilon$ for all $x, y \in D_n$ satisfies the hypothesis of Lemma 4. The following lemma achieves this goal.

Lemma 5. *Let $\varepsilon \geq 0$ and let $g : D_{2n} \to \mathbb{R}$ be such that for all $x, y \in D_n$,*

$$|g(x+y) - g(x) - g(y)| \leq \varepsilon.$$

Then, the function $h : \mathbb{Z} \to \mathbb{R}$ such that $h(x) = g(r_x) + q_x g(n)$ satisfies that for all $x \in \mathbb{Z}$,

$$|h(2x) - 2h(x)| \leq 2\varepsilon.$$

Proof. Let $x, y \in \mathbb{Z}$. By definition of h and since $r_{2x} = 2r_x - n(q_{2x} - 2q_x)$,

$$|h(2x) - 2h(x)| = |g(2r_x - n(q_{2x} - 2q_x)) - 2g(r_x) + (q_{2x} - 2q_x)g(n)|.$$

We will show that the right hand side of this equality is upper bounded by 2ε. Note that $q_{2x} - 2q_x \in \{-1, 0, 1\}$. We consider three cases depending on the value that this latter quantity takes.

CASE 1: Assume $q_{2x} - 2q_x = 0$. Then, since $r_x \in D_n$, the hypothesis implies that $|h(2x) - 2h(x)| = |g(2r_x) - 2g(r_x)| \leq \varepsilon$.

CASE 2: Assume now that $q_{2x} - 2q_x = 1$. Hence, $r_{2x} = 2r_x - n$ and

$$|h(2x) - 2h(x)| = |g(2r_x - n) - 2g(r_x) + g(n)|$$
$$\leq |g(2r_x) - 2g(r_x)| + |g(2r_x - n) + g(n) - g(2r_x)|$$
$$\leq 2\varepsilon,$$

where the first inequality is due to the triangle inequality and the second inequality follows from the hypothesis since $r_x, r_{2x} = 2r_x - n, n \in D_n$.

CASE 3: Assume $q_{2x} - 2q_x = -1$. Hence, $r_{2x} = 2r_x + n$ which is at most n. Thus, r_x cannot be positive. This implies that $r_x + n \in D_n$ and

$$
\begin{aligned}
|h(2x) - 2h(x)| &= |g(2r_x + n) - 2g(r_x) - g(n)| \\
&\leq |g(2r_x + n) - g(r_x + n) - g(r_x)| + |g(r_x + n) - g(r_x) - g(n)| \\
&\leq 2\varepsilon,
\end{aligned}
$$

where the first inequality is due to the triangle inequality and the second one follows from the hypothesis since $r_x + n, r_x, n \in D_n$.

An immediate consequence of the two previous results is the following:

Theorem 11. *Let $g : D_{2n} \to \mathbb{R}$ be a function such that for all $x, y \in D_n$,*

$$
|g(x + y) - g(x) - g(y)| \leq \varepsilon.
$$

Then, the linear function $l : D_{2n} \to \mathbb{R}$ defined by $l(n) = g(n)$ satisfies, for all $x \in D_n$,

$$
|g(x) - l(x)| \leq 2\varepsilon.
$$

4.4   Robustness

The results presented in the two previous sections yield the following:

Theorem 12. *Let $\varepsilon \geq 0$ and $0 \leq \eta < 1/96$ be constants, and let $P : D_{8n} \to \mathbb{R}$ be an application such that*

$$
\mathrm{Pr}_{x,y \in D_{4n}} \left[|P(x + y) - P(x) - P(y)| > \varepsilon \right] \leq \eta.
$$

Then, there exists a linear function $l : D_n \to \mathbb{R}$ such that

$$
\mathrm{Pr}_{x \in D_n} \left[|l(x) - P(x)| > 13\varepsilon \right] \leq 16\eta.
$$

Proof. Direct consequence of Theorem 9 and Theorem 11.

This last result gives us the analog of Theorem 3 that we need in order to establish the robustness of the **Absolute error Linearity Test**.

4.5   Self-Testing with Absolute Error

We now put together all the different pieces of the analyses of previous sections and establish the existence of an approximate self–tester for linearity.

Corollary 6. *Let $\mathcal{C}$ be the set of real valued functions over D_{8n} and let $\mathcal{L} \subseteq \mathcal{C}$ be the set of linear functions. Let $\eta > 0$ and $\varepsilon \geq 0$ be two constants. Then, there exists a $(D_{8n}, \varepsilon, \eta; D_n, 39\varepsilon, 577\eta)$–self–tester for $\mathcal{L}$ on $\mathcal{C}$ which uses for every confidence parameter $0 < \gamma < 1$, $O(\ln(1/\gamma)/\eta)$ calls to the oracle program, additions, comparisons, counter increments, and binary shifts.*

Proof. Consider the approximate test induced by the functional $\Phi(P, x, y) = P(x + y) - P(x) - P(y)$ where x and y are in D_{4n} and where the test error is 3ε. This approximate test clearly characterizes the family of linear functions, in fact, it gives rise to the **Absolute error Linearity Test**. Hence, by Lemma 3, it is $(\eta, 6\eta)$–continuous on D_{8n} with respect to the computational error term ε. Moreover, by Theorem 12, it is also $(96\eta', \eta')$–robust on D_n with respect to the computational error term $13(3\varepsilon)$. Therefore, Theorem 8 implies the desired result by fixing $\eta' = (6 + 1/96)\eta = 577\eta/96$.

4.6   Self-Correcting with Absolute Error

An obvious generalization of Definition 5 for any computational error term is:

Definition 14 (Approximate self–corrector). *Let $\mathcal{F} \subseteq \mathcal{C}$ be a function family from D to R and $D' \subseteq D$. Let $0 \leq \eta < 1$ and let ε, ε' be computational error terms for $\mathcal{C}$. An $(\eta, D, \varepsilon, D', \varepsilon')$–(approximate) self–corrector for $\mathcal{F}$ on $\mathcal{C}$ is a probabilistic oracle Turing machine T such that for every $P \in \mathcal{C}$, if $\mathsf{Dist}(P, f, D, \varepsilon) \leq \eta$ for some $f \in \mathcal{F}$, then for every $x \in D'$ and for every confidence parameter $0 < \gamma < 1$,*

$$\Pr\left[|T^P(x, \gamma) - f(x)| < \varepsilon'\right] > 1 - \gamma,$$

where the probability is taken over the internal coin tosses of T.

Of course, the above definition would be vacuous if we could not exhibit an example that satisfies it. We believe that in the same way that the majority argument gave rise to self–correctors, the median argument gives rise to approximate self–correctors. Below, we exhibit some supporting evidence for this claim by analyzing the problem of approximate self–correction of, the benchmark, class of linear functions.

Theorem 13. *Let $\mathcal{C}$ be the family of all real valued functions over D_{2n} and let $\mathcal{L} \subseteq \mathcal{C}$ be the set of linear functions. Then, for every $0 \leq \eta < 1/4$ and $\varepsilon \geq 0$, there is an $(\eta, D_{2n}, \varepsilon, D_n, 2\varepsilon)$–self–corrector for $\mathcal{L}$ on $\mathcal{C}$ which uses for every confidence parameter $0 < \gamma < 1$, $O(\ln(1/\gamma)/(1 - 4\eta)^2)$ calls to the oracle program, additions, comparisons, and counter increments.*

Proof. For some fixed $P \in \mathcal{C}$ assume $l : D_{2n} \to \mathbb{R}$ is a linear function such that $\mathsf{Dist}(P, l, D_{2n}, \varepsilon) \leq \eta$. Let N be a positive integer whose value will be determined later. Let T be the probabilistic Turing machine that on input $x \in D_n$, a constant $0 < \gamma < 1$, and oracle access to P, randomly chooses $y_1, \ldots, y_N \in D_n$ and then outputs $\mathsf{Med}_{i=1,\ldots,N} (P(x + y_i) - P(y_i))$. Note that,

$$\Pr\left[|T^P(x, \gamma) - l(x)| > 2\varepsilon\right]$$
$$= \Pr_{y_1,\ldots,y_N \in D_n} \left[|\mathsf{Med}_{i=1,\ldots,N} (P(x + y_i) - P(y_i)) - l(x)| > 2\varepsilon\right]$$
$$\leq \Pr_{y_1,\ldots,y_N \in D_n} \left[\mathsf{Med}_{i=1,\ldots,N} (|P(y_i) - l(y_i)|) > \varepsilon\right]$$
$$\quad + \Pr_{y_1,\ldots,y_N \in D_n} \left[\mathsf{Med}_{i=1,\ldots,N} (|P(x + y_i) - l(x + y_i)|) > \varepsilon\right]$$

$$\leq \mathsf{Pr}_{y_1,\dots,y_N \in D_n} \left[|\{y_i : |P(y_i) - l(y_i)| > \varepsilon\}| \geq \frac{N}{2} \right]$$

$$+ \mathsf{Pr}_{y_1,\dots,y_N \in D_n} \left[|\{y_i : |P(x + y_i) - l(x + y_i)| > \varepsilon\}| \geq \frac{N}{2} \right].$$

The Halving principle implies that both $\mathsf{Pr}_{y \in D_n} [|P(x + y) - l(x + y)| > \varepsilon]$ and $\mathsf{Pr}_{y \in D_n} [|P(y) - l(y)| > \varepsilon]$ are at most $2\mathsf{Dist}(P, l, D_{2n}, \varepsilon)$. Hence, a standard Chernoff bound yields the desired result when $N = \Omega(\ln(1/\gamma)(1 - 4\eta)^2)$.

4.7   Self-Testing a Specific Function with Absolute Error

One might expect that self–testing whether a program computes a specific linear function over $D_n \subseteq \mathbb{Z}$ would be achieved by replacing in the **Specific Linear Function Test** the **Linearity Test** by the **Absolute error Linearity Test**, equalities by approximate equalities, and a generator of $\mathbb{Z}$ (i.e., -1 or 1) by any non–zero $x \in D_n$. We shall see that one has to be more careful in the choice of the latter element. In particular one has to perform the following:

Absolute error Specific Linear Function Test(P, ε)
 – **Absolute error Linearity Test**(P, ε)
   1. Randomly choose $x, y \in D_{4n}$.
   2. Reject if $|P(x + y) - P(x) - P(y)| > \varepsilon$.
 – **Generator Test**(P)
   1. Randomly choose $x \in D_n$.
   2. Reject if $|P(x + n) - P(x) - f(n)| > \varepsilon$.

We now explain why in the second stage of the previous test the comparison between f and the self–corrected value of P is performed at n. First, recall that when the probability η of rejection by the **Absolute error Linearity Test** is low we have a guarantee that P is close to a linear function. A careful inspection of the proof of Theorem 12 elicits that when l is the real valued linear function defined over D_n that at n takes the value $\mathsf{Med}_{y \in D_{2n}} (P(n + y) - P(y))$,

$$\mathsf{Pr}_{x \in D_n} [|l(x) - P(x)| > 13\varepsilon] \leq 16\eta.$$

This justifies the comparison that is performed, in the second part of the above described approximate test, between $f(n)$ and the estimation $P(x + n) - P(n)$ of $l(n)$'s value. Lemma 3 and the following result yield the $(\eta, 22\eta)$–continuity of the **Absolute error Specific Linear Function Test** (when the test error is 5ε) on D_{8n} with respect to the computational error term ε.

Lemma 6. *Let $\varepsilon \geq 0$. Let P, f be real valued functions over D_{8n} such that f is linear. Then,*

$$\mathsf{Pr}_{x \in D_n} [|P(x + n) - P(x) - f(n)| > 2\varepsilon] \leq 16\mathsf{Pr}_{x \in D_{8n}} [|P(x) - f(x)| > \varepsilon].$$

Proof. The linearity of f, the union bound, and the Halving principle imply that

$$\Pr_{x\in D_n}\left[|P(x+n) - P(x) - f(n)| > 2\varepsilon\right]$$
$$\leq \Pr_{x\in D_n}\left[|P(x+n) - f(x+n)| > \varepsilon\right] + \Pr_{x\in D_n}\left[|P(x) - f(x)| > \varepsilon\right]$$
$$\leq 16\Pr_{x\in D_{8n}}\left[|P(x) - f(x)| > \varepsilon\right].$$

The following result implies the $(49\eta, \eta)$–robustness of the **Absolute error Specific Linear Function Test** (when the test error is ε) on D_n with respect to the computational error term 16ε.

Lemma 7. *Let $\varepsilon \geq 0$ and $0 \leq \eta < 1/96$ be constants and let $P, f : D_{8n} \to \mathbb{R}$ be mappings such that f is linear and the probability that the **Absolute error Specific Linear Function Test** rejects is at most η. Then,*

$$\Pr_{x\in D_n}\left[|f(x) - P(x)| > 16\varepsilon\right] \leq 49\eta.$$

Proof. Let $l : D_n \to \mathbb{R}$ be such that $l(n) = \mathsf{Med}_{y\in D_{2n}}\left(P(n+y) - P(y)\right)$ and linear. Implicit in the proof of Theorem 12 is that $\Pr_{x\in D_n}\left[|l(x) - P(x)| > 13\varepsilon\right] \leq 16\eta$ and that (see (2))

$$\Pr_{y\in D_n}\left[|g(n) - (P(n+y) - P(y))| > 2\varepsilon\right] < 32\eta.$$

Thus, $\Pr_{x\in D_n}\left[|f(x) - P(x)| > 16\varepsilon\right]$ is at most

$$\Pr_{x\in D_n}\left[|l(x) - P(x)| > 13\varepsilon\right] + \Pr_{x\in D_n}\left[\frac{|x|}{n}|P(x+n) - P(n) - l(n)| > 2\varepsilon\right]$$
$$+\Pr_{x\in D_n}\left[\frac{|x|}{n}|f(n) - P(x+n) + P(n)| > \varepsilon\right].$$

Since $|x|/n \leq 1$, the observation made at the beginning of this proof shows that the first and second term of the previous summation are bounded by 16η and 32η respectively. By the hypothesis and since $|x|/n \leq 1$, the last term of the summation is bounded by η.

Corollary 7. *Let C be the collection of all real valued functions over D_{8n} and let $f \in C$ be linear. Then, there exists a $(D_{8n}, \varepsilon, \eta; D_n, 80\varepsilon, 2113\eta)$–self–tester for f on C which uses for every confidence parameter $0 < \gamma < 1$, $O(\ln(1/\gamma)/\eta)$ calls to the oracle program, additions, comparisons, counter increments, and binary shifts.*

4.8   Beyond Self-Testing Approximate Linearity

So far we have discussed the approximate testing problem only for linear functions. The arguments we have used are also useful in testing non–linear functions. Nevertheless, a couple of new issues arise. To describe them we consider the problem of approximate testing whether a real valued function behaves like a degree d polynomial.

The characterization of degree d polynomials of Theorem 6, i.e., $\nabla_t^{d+1} f(x) = 0$ for all $x, t \in \mathbb{Z}_p$, still holds when $\mathbb{Z}_p$ is replaced by $\mathbb{Z}$. Hence, our discussion concerning approximate tests for class functions defined through functional equations suggest performing the following approximate test:

Absolute error Degree d Polynomial Test(P, ε)
1. Randomly choose $x \in D_m$ and $t \in D_n$.
2. Reject if $|\nabla_t^{d+1} P(x)| > \varepsilon$.

The above described approximate test was proposed and analyzed by Ergün, Kumar and Rubinfeld in [EKR96]. Since the approximate test with test error $2^{d+1}\varepsilon$ is $(d+2)$–local it is easily seen to be $(\eta, 2(d+2)\eta)$–continuous with respect to the computational error term ε, specifically:

Lemma 8. *Let $\varepsilon \geq 0$. Let P, Q be real valued functions over $D_{2(2d+3)n}$ such that Q is a polynomial of degree d. Then,*

$$\mathsf{Pr}_{x \in D_{(2d+3)n}, t \in D_n} \left[|\nabla_t^{d+1} P(x)| > 2^{d+1}\varepsilon \right] \leq$$
$$2(d+2)\mathsf{Pr}_{x \in D_{2(2d+3)n}} \left[|P(x) - Q(x)| > \varepsilon \right].$$

Proof. Simply observe that since Q is a degree d polynomial, $\nabla_t^{d+1} Q(x) = 0$ for all $x \in D_{(2d+3)n}$ and $t \in D_n$. Moreover, $\nabla_t^{d+1} = \sum_{k=0}^{d+1} (-1)^{d+1-k} \binom{d+1}{k} \nabla_{kt}$, $\sum_{k=0}^{d+1} \binom{d+1}{k} = 2^{d+1}$, and $|\nabla_t^{d+1} P(x)| > 2^{d+1}\varepsilon$ imply that $|P(x+it) - Q(x+it)| > \varepsilon$ for some $i \in \{0, \dots, d+1\}$. By the Halving principle, the probability that any of the latter events occur when $x \in D_{(2d+3)n}$ and $t \in D_n$ are randomly chosen is at most $2\mathsf{Pr}_{x \in D_{2(2d+3)n}} \left[|P(x) - Q(x)| > \varepsilon \right]$.

The approximate robustness of the **Absolute error Degree d Polynomial Test** is a consequence of the following:

Lemma 9. *Let $\varepsilon \geq 0$ and $0 \leq \eta < 1/(16(d+1)(d+2)^2)$ be constants, and let P be a real valued function defined over $D_{2(2d+3)n}$ such that*

$$\mathsf{Pr}_{x \in D_{(2d+3)n}, t \in D_n} \left[|\nabla_t^{d+1} P(x)| > \varepsilon \right] \leq \eta.$$

Then, there exists a function $g : D'_n \subseteq D_{(d+2)n} \to \mathbb{R}$ such that

$$\mathsf{Pr}_{x \in D'_n} \left[|g(x) - P(x)| > \varepsilon \right] \leq 2 \frac{|D_{(2d+3)n}|}{|D'_n|} \eta,$$

and for all $x, t \in D_n$,

$$|\nabla_t^{d+1} g(x)| \leq 4(2^{d+1} - 1)^2 \varepsilon.$$

Proof (Sketch). The proof is based on the median argument and follows the proof idea of Theorem 9. The choice of g is the same as in Theorem 7 but now $\mathsf{Maj}\,(\cdot)$ is replaced by $\mathsf{Med}\,(\cdot)$ and $\mathbb{Z}_p$ by D_n, i.e.,

$$g(x) = \mathsf{Med}_{t \in D_n} \left(\sum_{i=1}^{d+1} (-1)^{i+1} \binom{d+1}{i} P(x+it) \right).$$

As usual, approximate robustness leaves us with the need of a stability type result in order to establish robustness, in this case of the **Absolute error Degree d Polynomial Test**. We now undertake this endeavor.

For $t \in \mathbb{Z}^d$ denote by ∇_t the operator corresponding to the applications of $\nabla_{t_1}, \ldots, \nabla_{t_d}$. To avoid getting bogged down in technicalities and focus on the new issues that arise in approximate testing of non–linear function, we henceforth state the results for general d and restrict the proofs to the $d = 1$ case, i.e., the case of testing affine functions.

Lemma 10. Let $\varepsilon \geq 0$. Let $f : D_{(d+1)n} \to \mathbb{R}$ be such for all $t \in (D_n)^{d+1}$,

$$|\nabla_t f(0)| \leq \varepsilon,$$

Then, there exists a polynomial $h_d : D_n \to \mathbb{R}$ of degree at most d such that for all $x \in D_n$,

$$|f(x) - h_d(x)| \leq 2^d \left(\prod_{i=1}^{d} (2i - 1) \right) \varepsilon \leq 2^{2d} d! \varepsilon.$$

Proof (Sketch). We consider only the case of affine functions, i.e., $d = 1$. Let $G(t) = \nabla_t f(0) = f(t) - f(0)$. Then, for all $t_1, t_2 \in D_n$,

$$|G(t_1 + t_2) - G(t_1) - G(t_2)| = |\nabla_{t_1, t_2} f(0)| \leq \varepsilon.$$

Therefore, Theorem 11 implies that there exists a real valued linear function H over D_n such that $|G(t) - H(t)| \leq 2\varepsilon$ for all $t \in D_n$. Extending H linearly to all of $\mathbb{Z}$, defining f' over D_n by $f'(x) = f(x) - H(x)$, and observing that $H(0) = 0$ since H is linear, we get that for all $t \in D_n$,

$$|\nabla_t f'(0)| = |G(t) - H(t)| \leq 2\varepsilon.$$

To conclude, let $h(x) = f(0) + H(x)$ for all $x \in D_n$, and observe that h is an affine function such that $|f(x) - h(x)| = |\nabla_x f'(0)| \leq 2\varepsilon$.

Remark 2. For the case of general d, the proof of Lemma 10 has to be modified. First, G is defined for every $t \in \mathbb{Z}^d$ where it makes sense as $G(t) = \nabla_t f(0)$. Instead of Theorem 11, one needs a stability type result asserting the existence of a multi–linear function H on d variables which is close to G. Instead of a linear extension of H one relies on a multi–linear extension of H to $\mathbb{Z}^d$. The rest of the proof follows the same argument and exploits the fact that if $H'(x) = H(x, \ldots, x)$, then $\nabla_t H'(0) = d! H(t)$ for all $t \in \mathbb{Z}^d$.

We are not yet done proving the stability result we seek. Indeed, the conclusion of Lemma 9 is that $|\nabla_t^{d+1} g(x)|$ is bounded when $x, t \in D_n$. In contrast, the hypothesis of Lemma 10 requires a bound on $|\nabla_t g(0)|$ when $t \in (D_n)^{d+1}$. The following result links both bounds. But, the linkage is achieved at a cost. Indeed, although our assumption will be that $|\nabla_t^{d+1} g(x)|$ is bounded for a very large range of values of $x, t \in \mathbb{Z}$, our conclusion will be that $|\nabla_t g(0)|$ is bounded for a coarse range of values of $t \in \mathbb{Z}^{d+1}$.

Lemma 11. *Let $\varepsilon \geq 0$, $\mu_{d+1} = \mathrm{lcm}\{1, \ldots, d+1\}$, $m = \mu_{d+1}(d+1)n$, and g be a real valued function over $D_{(d+2)m}$. Let $f : D_{(d+1)n} \to \mathbb{R}$ be such that $f(x) = g(\mu_{d+1} \cdot x)$. If for all $x, t \in D_m$,*

$$|\nabla_t^{d+1} g(x)| \leq \frac{\varepsilon}{2^{d+1}},$$

then for all $\mathbf{t} \in (D_n)^{d+1}$,

$$|\nabla_{\mathbf{t}} f(0)| \leq \varepsilon.$$

Proof (Sketch). We consider only the case of affine functions, i.e., $d = 1$. Observe that

$$\nabla_{t_1, t_2} f(0) = \nabla_0^2 f(0) - \nabla_{-t_1}^2 f(t_1) - \nabla_{-t_2/2}^2 f(t_2) + \nabla_{-t_1 - t_2/2}^2 f(t_1 + t_2)$$

$$= \nabla_0^2 g(0) - \nabla_{-2t_1}^2 g(t_1) - \nabla_{-t_2}^2 g(t_2) + \nabla_{-2t_1 - t_2}^2 g(t_1 + t_2).$$

By hypothesis, each of the four terms in the last summation is upper bounded (in absolute value) by $\varepsilon/4$. The desired conclusion follows by triangle inequality.

Putting together Lemma 9, Lemma 10, and Lemma 11 one obtains the following result from which the $(2^{O(d \log d)} \eta, \eta)$–robustness with respect to the computational error term $2^{O(d \log d)} \varepsilon$ of the **Absolute error Degree d Polynomial Test** with test error ε immediately follows:

Theorem 14. *Let $\varepsilon \geq 0$, $\eta \geq 0$, $\mu_{d+1} = \mathrm{lcm}\{1, \ldots, d+1\}$, $m = \mu_{d+1}(d+1)n$, and let $kD_n = \{kx \in \mathbb{Z} : x \in D_n\}$ for any positive integer k. Let $P : D_{2(2d+3)m} \to \mathbb{R}$ be such that*

$$\Pr_{x \in D_{(2d+3)m}, t \in D_m} \left[|\nabla_t^{d+1} P(x)| > \varepsilon \right] \leq \eta.$$

Then, there exists a polynomial $h_d : \mu_{d+1} D_n \to \mathbb{R}$ of degree at most d such that

$$\Pr_{x \in \mu_{d+1} D_n} \left[|P(x) - h_d(x)| > 32^{d+1} d! \varepsilon \right] \leq 4(d+2)^2 \mu_{d+1} \eta.$$

Corollary 8. *Let $\varepsilon \geq 0$ and $\eta > 0$ be constants, $\mu_{d+1} = \mathrm{lcm}\{1, \ldots, d+1\}$, and $m = \mu_{d+1}(d+1)n$. Let $\mathcal{C}$ be the set of real valued functions over $D_{2(2d+3)m}$, and let $\mathcal{P}_d \subseteq \mathcal{C}$ be the set of degree d polynomials. Then, there exists a $(D_{2(2d+3)m}, \varepsilon, \eta; \mu_{d+1} D_n, 2^{O(d \log d)} \varepsilon, 2^{O(d \log d)} \eta)$–self–tester for $\mathcal{P}_d$ on $\mathcal{C}$ which uses for every confidence parameter $0 < \gamma < 1$, $O(\ln(1/\gamma)/\eta)$ calls to the oracle program, additions, comparisons, counter increments, and binary shifts.*

Proof (Sketch). Similar to the proof of Corollary 6 but now based on Lemma 8 and Theorem 14.

Note how the probability bounds in the statement of Theorem 14 depend exponentially in d. It is not clear that there has to be a dependency in d at all. A similar result without any dependency on d would be interesting. Even a polynomial in d dependency would be progress.

5   Testing with Error Depending on Input

In the preceding section we built self–testers for different function classes and domains for the case of absolute error. These self–testers exhibit the following characteristic: when the computational error term is a small constant they reject good programs, e.g, those in which the error in the computation of $P(x)$ grows with the size of x. If on the contrary, the computational error term is a large constant, they might pass programs that make incorrectly large errors in the computation of $P(x)$ for small values of x. In the next section we address the problem of self–testing when the computational error term can be proportional to the function value to be computed. In this section, we consider the intermediate case where the computational error terms are measured relative to some pre-specified function of the input x to the program P being tested. In particular, they do not depend on the function f purportedly being computed. The results presented here appeared in [KMS99].

In order to achieve the above stated goal, we generalize the arguments discussed in the preceding sections. We begin by pointing out that a careful inspection of the proofs of Theorem 9 and Theorem 11 yield that they still hold as long as the test error satisfies a collection of properties captured by the following:

Definition 15 (Valid error terms of degree $p \in \mathbb{R}$). *These are nonnegative functions $\beta : \mathbb{Z} \times \mathbb{Z} \to \mathbb{R}^+$ which are, in each of their coordinates, even and nondecreasing for nonnegative integers, and such that $\beta(2s, 2t) \leq 2^p \beta(s,t)$ for all integers s, t.*

Examples of valid error terms of degree p are $\beta(s,t) = |s|^p + |t|^p$ and $\beta(s,t) = \mathrm{Max}\{c, |s|^p, |t|^p\}$ for some nonnegative real constant c. Whenever it is clear from context, we abuse notation and interpret a degree p error $\beta(\cdot, \cdot)$ as the function of one variable, denoted $\beta(z)$, that evaluates to $\beta(z, z)$ at z. Also, for $0 \leq p < 1$, we set $C_p = (1 + 2^p)/(2 - 2^p)$ and henceforth throughout this section use this notation.

When it is clear from the context, speaking about valid error terms β will both refer to test errors and computational error terms of the form $\varepsilon(x, v) = \beta(x)$.

5.1   Stability

By our choice of definition for test error depending on input size, with some effort but no new ideas, one can generalize the proof arguments of Lemma 4 and Lemma 5 and derive the following analog of Theorem 11:

Theorem 15. *Let $\beta(\cdot, \cdot)$ be a valid error term of degree p where $0 \leq p < 1$. Let $g : D_{2n} \to \mathbb{R}$ be such that for all $x, y \in D_n$,*

$$|g(x + y) - g(x) - g(y)| \leq \beta(x, y).$$

Then, the linear mapping $T : \mathbb{Z} \to \mathbb{R}$ defined by $T(n) = g(n)$ is such that for all $x \in D_n$,

$$|g(x) - T(x)| \leq C_p \beta(x).$$

This last theorem is the stability type result we need to establish robustness once we prove the approximate robustness of the analog of the **Absolute error Linearity Test** where instead of comparing $|P(x+y) - P(x) - P(y)|$ to a fixed constant ε the comparison is made against $\beta(x, y)$.

5.2   Approximate Robustness

We again rely on the median argument, but there is a crucial twist that needs to be introduced in order to address the cases of non–constant test errors we are concerned with. To explain the new twist, recall that in the median argument one begins by defining a function g whose value at x is the median of a multiset S_x whose elements depend on x and P, i.e.,

$$g(x) = \mathsf{Med}_{s \in S_x}(s).$$

Each value s in S_x is seen as an estimation of the correct value that P takes on x. We would like $g(x)$ to be a very good estimation of the correct value taken by P on x. But now, how good an estimation is depends on the size of x. The smaller the size of x, the more accurate we want the estimation to be. This forces a new definition for $g(x)$, specially when x is small. The following result illustrates this point for the case of linearity testing with valid error terms.

Theorem 16. *For $0 \le \delta \le 1$ and a valid error term $\beta(\cdot, \cdot)$ of degree $0 \le p < 1$ define $\widetilde{\beta}(z) = \beta(\mathrm{Max}\{n\sqrt{\delta}, |z|\})$. Let $P : D_{8n} \to \mathbb{R}$ be a mapping such that*

$$\mathsf{Pr}_{x,y \in D_{4n}}\left[|P(x+y) - P(x) - P(y)| > \beta(x, y)\right] \le \delta/384.$$

Then, there exists a function $g : D_{2n} \to \mathbb{R}$ such that

$$\mathsf{Pr}_{x \in D_n}\left[|g(x) - P(x)| > \widetilde{\beta}(x)\right] \le \delta/6,$$

and for all $a, b \in D_n$,

$$|g(a+b) - g(a) - g(b)| \le 16\,\mathrm{Max}\{\widetilde{\beta}(a), \widetilde{\beta}(b)\}.$$

Proof (Sketch). The key point is the choice of g, i.e., for $x \in D_n$ define

$$g(x) = \begin{cases} \mathsf{Med}_{y \in D_{|x|}}\left(P(x+y) - P(y)\right), & \text{if } |x| \ge n\sqrt{\delta}, \\[2mm] \mathsf{Med}_{y \in D_{n\sqrt{\delta}}}\left(P(x+y) - P(y)\right), & \text{otherwise.} \end{cases}$$

Then, following the proof argument of Theorem 9, one obtains the desired conclusion (although not without effort).

Note how in the above definition of g, the median is taken over sets of different sizes. We henceforth refer to this variation of the median argument as the variable size median argument.

5.3   Robustness

The main goal of the two previous sections was to help establish the following:

Theorem 17. *Let $0 \leq \delta \leq 1$ and $\beta(\cdot, \cdot)$ be a valid error term of degree $0 \leq p < 1$. If $P : D_{8n} \to \mathbb{R}$ is such that*

$$\Pr_{x,y \in D_{4n}} \left[|P(x+y) - P(x) - P(y)| > \beta(x, y) \right] \leq \delta/384,$$

then there exists a linear function $T : \mathbb{Z} \to \mathbb{R}$ such that

$$\Pr_{x \in D_n} \left[|P(x) - T(x)| > 17 C_p \beta(x) \right] \leq 7\sqrt{\delta}/6.$$

Proof. Let $\widetilde{\beta}(z) = \beta(\mathrm{Max}\{n\sqrt{\delta}, |z|\})$ and $\beta'(x, y) = 16 \, \mathrm{Max}\{\widetilde{\beta}(x), \widetilde{\beta}(y)\}$. Since $\beta'(\cdot, \cdot)$ is a valid error term of degree p, Theorem 15 and Theorem 16 imply that,

- there is a function $g : D_{2n} \to \mathbb{R}$ such that when $x \in D_n$ is randomly chosen, $|g(x) - P(x)| > \widetilde{\beta}(x)$ with probability at most $\delta/6$, and
- there is a linear map $T : \mathbb{Z} \to \mathbb{R}$ such that $|g(x) - T(x)| \leq 16 C_p \widetilde{\beta}(x)$ for all $x \in D_n$.

Since $1 \leq C_p$, if $|P(x) - T(x)| > 17 C_p \widetilde{\beta}(x)$, then $|g(x) - P(x)| > \widetilde{\beta}(x)$ when $x \in D_n$. Hence, $\Pr_{x \in D_n} \left[|g(x) - P(x)| > 17 C_p \widetilde{\beta}(x) \right]$ is at most $\delta/6 \leq \sqrt{\delta}/6$. To conclude the proof observe that $\widetilde{\beta}(x) = \beta(x)$ with probability at least $1 - \sqrt{\delta}$ when x is randomly chosen in D_n.

The previous result is the analog of Theorem 12 one needs to construct an approximate self–tester for linear functions provided the valid error term $\beta(\cdot, \cdot)$ is easily computable. Indeed, given oracle access to the program P and a valid error term $\beta(\cdot, \cdot)$, one can perform the following procedure:

---
1. Randomly choose $x, y \in D_{4n}$.
2. Reject if $|P(x+y) - P(x) - P(y)| > \beta(x, y)$.

We are now faced with a crucial difference between testing in the absolute error case and the case where the test errors depend on the size of the inputs. The point is that the above defined approximate test can be implemented provided one has a way of computing efficiently the valid error term, i.e., $\beta(\cdot, \cdot)$. Moreover, we would certainly like that computing the valid error term is simpler than computing whatever function P purportedly computes. In the case of linearity testing this is not always the case if the valid error term is a non–linear function, say $\beta(x, y) = \sqrt{|x|} + \sqrt{|y|}$. It is interesting to note that in most of the testing literature it is implicitly assumed that the test error is efficiently computable (always 0 in the case of exact testing and a fixed constant hardwired into the testing programs in the case of testing with absolute error). Fortunately, a good approximation of the test error suffices for self–testing. More precisely, provided the valid error term $\beta(\cdot, \cdot)$ is such that for some positive constants λ and λ' there is a function $\varphi(\cdot, \cdot)$ that is (λ, λ')–*equivalent to*

$\beta(\cdot,\cdot)$, i.e., $\lambda\varphi(s,t) \geq \beta(s,t) \geq \lambda'\varphi(s,t)$ for all integers s,t. In addition, one desires that evaluating φ is asymptotically faster than executing the program being tested, say it only requires additions, comparisons, counter increments, and binary shifts. Surprisingly, this is feasible. For example, let k and k' be positive integers and let $\lg(n)$ denote the length of an integer n in binary. (Note that $\lg(n) = \lceil \log_2(|n|+1) \rceil$ or equivalently $\lg(0) = 0$ and $\lg(n) = \lfloor \log_2(|n|) \rfloor + 1$ if $n \neq 0$.) Then, $\beta(s,t) = 2^{k'}(|s|^{1/2^k} + |t|^{1/2^k})$ or $\beta(s,t) = 2^{k'}\operatorname{Max}\{|s|^{1/2^k}, |t|^{1/2^k}\}$ are valid error terms of degree $1/2^k$ which are $(1, 1/2)$–equivalent to $\varphi(s,t) = 2^{k'}(2^{\lceil \lg(s)/2^k \rceil} + 2^{\lceil \lg(t)/2^k \rceil})$ and $\varphi(s,t) = 2^{k'+\operatorname{Max}\{\lceil \lg(s)/2^k \rceil, \lceil \lg(t)/2^k \rceil\}}$ respectively. The computation of these latter functions requires only counter increments and shifting bits.

We have finally arrived at a point where we can propose an approximate test for linearity in the case of valid error terms $\beta(\cdot,\cdot)$ of degree $0 \leq p < 1$ for which there exists an equivalent function $\varphi(\cdot,\cdot)$, i.e.,

---

Input Size Relative error Linearity Test(P, φ)
1. Randomly choose $x, y \in D_{4n}$.
2. Reject if $|P(x + y) - P(x) - P(y)| > \varphi(x, y)$.

5.4   Continuity

As usual, establishing continuity, in this case of the **Input Size Relative error Linearity Test** is simple. We had not done so before simply because we had no candidate test to analyze. Below we establish the $(\eta, 6\eta)$–continuity with respect to the computational error term β of the mentioned approximate test with test error $\beta/4$, but to succeed we need and additional condition on the valid error term. We say that a valid error term $\beta(\cdot,\cdot)$ is c–testable, where c is a constant, if $\beta(s) + \beta(t) + \beta(s+t) \leq c\beta(s,t)$ for all s and t. For example, for k and k' integers, k positive, $\beta(s,t) = 2^{k'}(|s|^{1/2^k} + |t|^{1/2^k})$ and $\beta(s,t) = 2^{k'}\operatorname{Max}\{|s|^{1/2^k}, |t|^{1/2^k}\}$ are 4–testable valid error terms.

Lemma 12. *Let $\beta(\cdot,\cdot)$ be a 4–testable valid error term. Let P, l be real valued functions over D_{8n} such that l is linear. Then,*

$$\operatorname{Pr}_{x,y\in D_{4n}}\left[|P(x+y) - P(x) - P(y)| > \beta(x,y)\right] \leq$$

$$6\operatorname{Pr}_{x\in D_{8n}}\left[|P(x) - l(x)| > \frac{\beta(x)}{4}\right].$$

Proof. Let $\beta' = \beta/4$. By the Halving principle,

$$\operatorname{Pr}_{x\in D_{4n}}\left[|P(x) - l(x)| > \beta'(x)\right] \leq 2\operatorname{Pr}_{z\in D_{8n}}\left[|P(z) - l(z)| > \beta'(z)\right],$$

$$\operatorname{Pr}_{y\in D_{4n}}\left[|P(y) - l(y)| > \beta'(y)\right] \leq 2\operatorname{Pr}_{z\in D_{8n}}\left[|P(z) - l(z)| > \beta'(z)\right],$$

$$\operatorname{Pr}_{x,y\in D_{4n}}\left[|P(x+y) - l(x+y)| > \beta'(x+y)\right] \leq 2\operatorname{Pr}_{z\in D_{8n}}\left[|P(z) - l(z)| > \beta'(z)\right].$$

Hence, since $\beta'(s) + \beta'(t) + \beta'(s+t) \leq \beta(s,t)$, the union bound implies the desired result.

5.5   Self-Testing with Error Relative to Input Size

We now piece together the results and concepts introduced in previous sections and establish the existence of realizable approximate self–testers for the case when the computational error term is allowed to depend on the size of the input. We stress that the existence of computationally efficient self–testers of this type is not a priori obvious since the test error might not be efficiently computable.

Theorem 18. *Let $0 < \eta \leq 1$ and $\beta(\cdot,\cdot)$ be a 4–testable valid error term of degree $0 < p < 1$ such that $\varphi(\cdot,\cdot)$ is (λ,λ')–equivalent to $\beta(\cdot,\cdot)$. Then, there is a $(D_{8n}, \beta/(4\lambda), \eta/384; D_n, 17C_p\beta/\lambda', 7\sqrt{\eta}/6)$–self-tester for the class of real valued linear functions over D_{8n}. Moreover, the self–tester uses for every confidence parameter $0 < \gamma < 1$, $O(\ln(1/\gamma)/\eta)$ calls to the oracle program, additions comparisons, counter increments, and binary shifts.*

Proof. First, assume that $\beta(\cdot,\cdot)$ is efficiently computable and consider the approximate test induced by the functional $\Phi(P, x, y) = P(x + y) - P(x) - P(y)$ where x and y are in D_{4n} and the test error is $\beta(\cdot,\cdot)$. This approximate test clearly characterizes the family of linear functions. In fact, it gives rise to the **Input Size Relative error Linearity Test**(P, β). Hence, by Lemma 12, it is $(\eta, 6\eta)$–continuous on D_{8n} with respect to the computational error term $\beta/4$. Moreover, by Theorem 17, it is also $(7\sqrt{\delta}/6, \delta/384)$–robust on D_n with respect to the computational error term $17C_p\beta$. Therefore, Theorem 8 implies the desired result by fixing $6\eta < \delta/384$.

To conclude the proof, we need to remove the assumption that the valid error term $\beta(\cdot,\cdot)$ is efficiently computable. To do so, consider the self–tester that performs sufficiently many independent rounds of the **Input Size Relative error Linearity Test**(P, φ). An analysis almost identical to the one described above applied to the new self–tester yields the desired result.

Remark 3. The $\sqrt{\eta}$ dependency in the previous theorem, which is inherited from Theorem 17 is not the type of probability bound one usually sees in the context of exact and absolute error self–testing. Nevertheless, as the example below shows, this dependency seems to be unavoidable in the case of testing with errors that depend on the size of the input.

Let n be a positive integer, $0 < p < 1$, $0 < \delta < 1/4$, $\theta, c > 0$, $\beta(x,y) = \theta \operatorname{Max}\{|x|^p, |y|^p\}$, and consider the function $P : \mathbb{Z} \to \mathbb{R}$ such that (see Fig. 3)

$$P(x) = \begin{cases} -\theta(n\sqrt{\delta})^p, & \text{if } -n\sqrt{\delta} \leq x < 0, \\ \theta(n\sqrt{\delta})^p, & \text{if } 0 < x \leq n\sqrt{\delta}, \\ 0, & \text{otherwise.} \end{cases}$$

Observe that if $|x|$ or $|y|$ is greater than $n\sqrt{\delta}$ then $|P(x+y) - P(x) - P(y)| \leq 2\beta(x,y)$. Hence, if $n' \geq n$, with probability at most δ it holds that $|P(x + y) - P(x) - P(y)| > 2\beta(x,y)$ when x and y are randomly chosen in $D_{n'}$.

One can show that for every linear function T, when $x \in D_n$ is randomly chosen, $|P(x) - T(x)| > c\beta(x)$ with probability greater than $\sqrt{\delta}/(2(\operatorname{Max}\{1, 2c\})^{1/p})$.

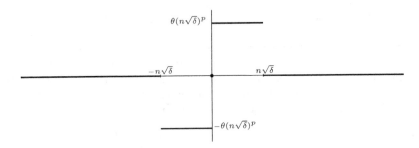

Fig. 3. The function P.

Similar results as those stated above for linear functions hold for the class of polynomials. Their derivation is based on the arguments discussed in the previous as well as this section. Unfortunately, these arguments give rise to technically complicated proofs (for details see [KMS99]). Simplifications are certainly desirable.

6   Testing with Relative Error

In this section we consider the testing problem in the case where the allowed computational test error is proportional to the (absolute value of) the correct output one wishes to compute, i.e., the so called case of relative error. Again, we have oracle access to a program P purportedly computing a function belonging to a class of functions $\mathcal{F}$. The specific function f which P purportedly computes is unknown if there is more than one element in $\mathcal{F}$. The accuracy we wish P to compute f on x depends on the unknown value of $f(x)$. Thus, it is not a priori clear than one can self–test in the context of relative error. The discussion we now undertake will establish the plausibility of this task.

The forthcoming presentation is based on [Mag00a]. It shows how to build a self–tester for the class of linear functions in the case of relative error. The construction proceeds in two stages. First, one builds a linearity self–tester for the linear error case, i.e., the case where the test error is a linear function of the (absolute vale of) the input. This self–tester is then modified so as to successfully handle the case of relative error. The linear and relative error case, although related, exhibit a crucial difference. In the former case, since the error is just a known constant times the (absolute value of) the input, one knows (and thus can compute) the test error. In the latter case, one can not directly evaluate the test error since the function purportedly being computed by the oracle program is unknown and the test error depends on this value.

Note that in the context of linearity testing over rational domains, relative errors are functions that map x to $\theta|x|$ where θ is some unknown positive constant. Even if θ was known, Theorem 17 would not be applicable since it does not hold when $p = 1$. To see this, consider the real valued function over $\mathbb{Z}$ defined by $f(x) = \theta x \log_2(1 + |x|)$ for some $\theta > 0$. In [RS92a], it is shown that

$|f(x+y) - f(x) - f(y)| \leq 2\theta \operatorname{Max}\{|x|, |y|\}$ for all $x, y \in \mathbb{Z}$. Clearly, no linear function is close to f. Hence, in the case of linear error, the **Input Size Relative error Linearity Test** is not a good self-tester for linearity. In order to overcome this situation it is natural to either use a different test error or modify the test. Based on the former option, in previous sections, approximate self-testers were derived from exact self-testers. In contrast, to derive approximate self-testers in the case of linear error the latter path is taken.

When x is large, say $|x| \geq n/2$, a linear error term is essentially an absolute error term. When x is small, say $|x| < n/2$, we would like to efficiently amplify the linear error term to an absolute one. This can be done by multiplying x by the smallest power of 2 such that the absolute value of the result is at least $n/2$. This procedure can be efficiently implemented by means of binary shifts. Formally, each x is multiplied by 2^{k_x} where

$$k_x = \operatorname{Min}\{k \in \mathbb{N} : 2^k |x| \geq n/2\}.$$

(See Fig. 4 for an example where $n/8 < x < n/4$.)

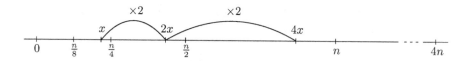

Fig. 4. Amplification procedure

The amplification procedures described above leads to the following new functional equation characterization of the class of linear functions (whose domain is D_{8n}):

$$\forall x, y \in D_{4n}, \qquad f(2^{k_x} x + y) - 2^{k_x} f(x) - f(y) = 0.$$

Note how this new characterization of linear functions relies not only on the additive properties of linear functions, but also on their homothetic properties.

6.1   Linear Error

The previous section's functional equation characterization of linear functions leads in the standard way to a functional equation test. Specifically, for $\theta \geq 0$ it yields the following:

> **Linear error Linearity Test(P, θ)**
> 1. Randomly choose $x, y \in D_{4n}$.
> 2. Reject if $|P(2^{k_x} x + y) - 2^{k_x} P(x) - P(y)| > \theta 2^{k_x} |x|$.

We henceforth denote by $\operatorname{Rej}(P, \theta)$ the rejection probability of P by the **Linear error Linearity Test**. The following claim establishes the continuity of this approximate test.

Lemma 13. *Let $\theta \geq 0$ and $\mathcal{L}$ be the set of linear functions over $\mathbb{Z}$. Then, for every $P : D_{8n} \to \mathbb{R}$,*

$$\mathsf{Rej}(P, \theta) \leq 6\mathsf{Dist}(P, \mathcal{L}, D_{8n}, \theta|x|/18).$$

The proof of robustness for the **Linear error Linearity Test** follows the usual two step approach where both the approximate robustness and the stability of the test are established. The first of these properties is guaranteed by the following:

Theorem 19. *Let $0 \leq \eta < 1/512$ and $\theta \geq 0$. Let $P : D_{8n} \to \mathbb{R}$ be such that*

$$\mathsf{Pr}_{x,y \in D_{4n}} \left[|P(2^{k_x}x + y) - 2^{k_x}P(x) - P(y)| > \theta 2^{k_x}|x| \right] \leq \eta.$$

Then, the function $g : D_{2n} \to \mathbb{R}$ defined by

$$g(x) = \frac{1}{2^{k_x}}\mathsf{Med}_{y \in D_{2n} : xy \geq 0}\left(P(2^{k_x}x + y) - P(y)\right),$$

is such that

$$\mathsf{Pr}_{x \in D_n}\left[|P(x) - g(x)| > \theta|x|\right] \leq 32\eta.$$

Moreover, $g(x) = g(2^{k_x}x)/2^{k_x}$ for all $x \in D_{2n}$, $|g(n) + g(-n)| \leq 16\theta n$, and for all x and y in $\{n/2, \dots, n\}$ (respectively $\{-n/2, \dots, -n\}$)

$$|g(x + y) - g(x) - g(y)| \leq 24\theta n.$$

Proof (Sketch). The proof follows the median function argument. The main difference is that we now have to cope with amplification terms. The closeness of g to P follows from the definition of g, the median principle, and the bound on rejection probability of the approximate test. The homethetic property of g under the amplification procedure follows directly from g's definition. It only remains to prove the approximate additivity of g in x and y (that is $g(x + y)$ is close to $g(x) + g(y)$) when the amplification terms of x, y and $x + y$ are all the same. More precisely, when both x and y belong to either $\{n/2, \dots, n\}$ or $\{-n/2, \dots, -n\}$ and when $\{x, y\} = \{-n, n\}$. This partly justifies the restriction on the set of elements of D_{2n} over which the median is taken (when $xy \geq 0$ one knows that the absolute values of $2^{k_x}x$, y, and $2^{k_x}x + y$ are all at least $n/2$). Therefore, they have no amplification factors associated to them.

Note that the approximate additivity of g over $\{n/2, \dots, n\}$ and $\{-n/2, \dots, -n\}$ established by the previous result guarantees, due to g's homothetic property, its approximate additivity over small elements of g's domain.

The stability of the **Linear error Linearity Test** is established by the following:

Theorem 20. *Let $\theta_1, \theta_2 \geq 0$. Let $g : D_{2n} \to \mathbb{R}$ be such that $g(x) = g(2^{k_x}x)/2^{k_x}$ for all $x \in D_{2n}$, $|g(n) + g(-n)| \leq \theta_1 n$, and for all x and y in $\{n/2, \dots, n\}$ (respectively $\{-n/2, \dots, -n\}$)*

$$|g(x + y) - g(x) - g(y)| \leq \theta_2 n.$$

Then, the linear function $l : D_n \to \mathbb{R}$ *defined by* $l(n) = g(n)$ *satisfies, for all* $x \in D_n$,

$$|g(x) - l(x)| \le (\theta_1 + 5\theta_2)|x|.$$

Proof (Sketch). The idea is to prove first that g is close to some linear function l (respectively l') on $\{n/2, \dots, n\}$ (respectively $\{-n/2, \dots, -n\}$), but in the absolute error sense. This can be achieved by an argument similar to the one used in the proof of Theorem 15. It follows that l and l' are necessarily close since $g(n)$ and $g(-n)$ are close to each other. Then, the homothetic property of g is used to transform absolute error bounds on the distance between g and l over $\{n/2, \dots, n\}$ and $\{-n/2, \dots, -n\}$, into linear error bounds over all of D_n.

Theorem 19 and Theorem 20 immediately yield:

Theorem 21. *Let* $\theta \ge 0$, $0 \le \eta \le 1/512$, $P : D_{8n} \to \mathbb{R}$, *and* $l : D_n \to \mathbb{R}$ *be the linear function such that*

$$l(n) = \mathsf{Med}_{y \in D_{2n}:y \ge 0} \left(P(n + y) - P(y) \right).$$

Then,

$$\mathsf{Rej}(P, \theta) \le \eta \implies \mathsf{Dist}(P, l, D_n, 137\theta|x|) \le 32\eta.$$

6.2   From Linear Error to Relative Error

We now undertake the second stage of the construction of the self–tester for the class of linear functions in the case of relative error. Specifically, we modify the **Linear error Linearity Test** so it can handle relative test errors. In order to explain this modification, consider a program P that approximately computes (with respect to relative errors) a linear function l. Then, one could allow a test error proportional to $l(n)$ in the **Linear error Linearity Test**. Since l is unknown, we need to estimate its value at n. Although P is close to l the value $P(n)$ might be very far from $l(n)$. Thus, $P(n)$ is not necessarily a good estimation of $l(n)$. We encountered a similar situation when self–testing a specific function. We addressed it via self–correction. The same approach succeeds here. This leads to the **Relative error Linearity Test** described below. To state it we first need to define (over $\mathbb{Z}$) the real valued function $ext(P, G)$ by:

$$ext(P, G)(x) = \begin{cases} P(x), & \text{if } x \in D_n, \\ ext(P, G)(x - n) + G, & \text{if } x > n, \\ ext(P, G)(x + n) - G, & \text{if } x < -n. \end{cases}$$

Then, the modified **Linear error Linearity Test** becomes the

---

Relative error Linearity Test(P, θ)
1. Randomly choose $y \in \{0, \dots, n\}$.
2. Compute $G_y = P(n - y) + P(y)$.
3. Compute $\tilde{\theta} = \theta|G_y|/n$.
4. Call **Linear error Linearity Test**$(ext(P, G_y), \tilde{\theta})$.

We henceforth denote by $\mathsf{Rej}^r(P, \theta)$ the rejection probability of P by the **Relative error Linearity Test** and let $\mathsf{Dist}^r(\cdot, \cdot, \cdot, \theta)$ denote $\mathsf{Dist}(\cdot, \cdot, \cdot, \varepsilon)$ when the computational error term ε is $\varepsilon(x, v) = \theta|v|$. The following results establish both the continuity and the robustness of the **Relative error Linearity Test**.

Lemma 14. *Let* $0 \leq \theta \leq 18$, $\mathcal{L}$ *be the set of linear functions over* $\mathbb{Z}$, *and* $P : D_n \to \mathbb{R}$. *Then,*

$$\mathsf{Rej}^r(P, \theta) \leq 10\mathsf{Dist}^r(P, \mathcal{L}, D_n, \theta/72).$$

Proof. Let $l : D_n \to \mathbb{R}$ be a linear function such that $\mathsf{Dist}^r(P, l, D_n, \theta/72) = \mathsf{Dist}^r(P, \mathcal{L}, D_n, \theta/72) = \eta$. For $y \in \{0, \dots, n\}$ let $G_y = P(n - y) + P(y)$, $\tilde{\theta} = \theta|G_y|/n$, and $\tilde{P}_y = ext(P, G_y)$. By Lemma 2, $|G_y - l(n)| \leq \theta|l(n)|/36$ with probability greater than $1 - 4\eta$ when y is randomly chosen in $\{0, \dots, n\}$. If this latter inequality is satisfied, then $\mathsf{Dist}^r(\tilde{P}_y, l, D_{8n}, \theta/36) \leq \eta$. Since $\theta/36 \leq 1/2$, the assumed inequality also implies that $|l(n)| \leq 2|G_y|$. Therefore, it follows that $\mathsf{Dist}(\tilde{P}_y, l, D_{8n}, \theta|x|/18) \leq \eta$. Lemma 13 implies that the rejection probability of the **Linear error Linearity Test**$(ext(P, G_y), \tilde{\theta})$ is at most 6η. It immediately follows that $\mathsf{Rej}^r(P, \theta) \leq (6 + 4)\eta$.

Theorem 22. *Let* $\theta \geq 0$, $0 \leq \eta \leq 1/512$, $\mathcal{L}$ *be the set of linear functions over* $\mathbb{Z}$, *and* $P : D_n \to \mathbb{R}$. *Then,*

$$\mathsf{Rej}^r(P, \theta) \leq \eta \Longrightarrow \mathsf{Dist}^r(P, \mathcal{L}, D_n, 137\theta) \leq 32\eta.$$

Proof. Assume $\mathsf{Rej}^r(P, \theta) \leq \eta$. Then, there exists a $y \in D_n$ such that for $G_y = P(n - y) + P(y)$ and $\tilde{\theta} = \theta|G_y|/n$, the rejection probability of **Linear error Linearity Test**$(ext(P, G_y), \tilde{\theta})$ is at most η. Thus, by Theorem 21, the linear function $l : D_n \to \mathbb{R}$ defined by

$$l(n) = \mathsf{Med}_{y \in D_{2n}: y \geq 0} \left(P(n + y) - P(y) \right),$$

is such that $\mathsf{Dist}(P, l, D_n, 137\tilde{\theta}|x|) \leq 32\eta$. Then, it must be that $l(n) = G_y$ and therefore

$$\mathsf{Dist}^r(P, l, D_n, 137\theta) = \mathsf{Dist}(P, l, D_n, 137\tilde{\theta}|x|) \leq 32\eta.$$

Similar results also hold for the class of multi–linear functions (see [Mag00b] for details).

7   Beyond Testing Algebraic Functions

Since the pioneering work of Blum et al. [BLR90] appeared the concepts and paradigms discussed so far in this survey have been extended and studied under different contexts. This has been done in order to widen the scope of applicability of the concepts and results obtained in the self–testing literature. Below we discuss some of these extensions and new scenarios.

7.1   Testing and Probabilistically Checkable Proofs

The results of [BFLS91,AS92b,ALM+92] concerning probabilistically checkable proofs (PCPs) enable the encoding of mathematical proofs so as to allow very efficient probabilistic verification. The latter consists of a simple randomized test that looks at a few bits of the proof and decides to accept or reject the proof's validity by performing a simple computation on those bits. Valid proofs are always accepted. Incorrect proofs are rejected with a non–negligible probability.

PCPs are built by recursion [AS92b]. Each level of the recursion uses a distinct form of error-correcting code. Correct encodings are viewed as representations of functions that satisfy a pre-specified property. Thus, a central problem in the construction of PCPs is to probabilistically check (test) whether a function satisfies a given property with as few queries as possible. Among the typical properties that come up in the PCP context are linearity [BGLR93,BS94,BCH+95, Tre98], multi–linearity [BFL90,FGL+91], low–individual degree [BFLS91,AS92b, PS94], low–total degree [ALM+92,AS97], and membership in the so called "long code" [Hås96,Hås97,Tre98]. Testing that a given function satisfies one of these properties is referred to as *low–degree testing*.

In the context of PCPs, thus in low–degree testing, the main concern is to minimize the number of oracle queries and the randomness used. It is not essential that the verification procedure be computationally more efficient than computing the functions of the class one wants to test. Also, continuity of the test is not so much of an issue, typically it only matters that the test never rejects an oracle that represents a function with the desired property. Robustness is the real issue.

Low–degree testing takes place in an adversarial scenario where the verification procedure is thought of as a verifier that has access to an oracle written down by a prover. The prover wishes to foul the verifier into believing the oracle is the table of a function satisfying a specific property. The verifier wants to determine whether this is true using as few probes and random bits as possible. To simplify his task the verifier may force the prover to add structure to the oracle. Moreover, he may choose a scenario where performing the verification is easier. For the sake of illustration and concreteness we shall consider below our benchmark linearity testing problem but in the PCP context.

The discussion that follows is taken from [KR97]. Assume G and H are finite abelian groups and $P : G \to H$. We want to verify whether P is linear, i.e., $P(x + y) = P(x) + P(y)$ for all $x, y \in G$. To simplify the verification procedure we choose G and H so they have a rich structure. Specifically, for a prime field $\mathbb{Z}_p$, we fix $G = \mathbb{Z}_p^n$ and $H = \mathbb{Z}_p$. Since $\mathbb{Z}_p$ is a prime field, P is linear if and only if $P(x) = \sum_{i=1}^n \alpha_i x_i$ for some $\alpha_1, \ldots, \alpha_n \in \mathbb{Z}_p$. For $x \in \mathbb{Z}_p^n \setminus \{0\}$, denote by L_x the line in $\mathbb{Z}_p^n$ passing through x and 0, i.e., $L_x = \{tx : t \in \mathbb{Z}_p\}$. Observe that if $L_x \neq L_y$, then $L_x \cap L_y = \{0\}$. Note also that every linear function l over $\mathbb{Z}_p^n$ is such that $l(tx) = tl(x)$ for all $t \in \mathbb{Z}_p$ and $x \in \mathbb{Z}_p^n$. Hence, knowing the value of l at any non–zero element of a line completely determines the value of l over that line. We take advantage of this fact to facilitate the verification task. Indeed, we ask the prover to write down, for each line $L \subseteq \mathbb{Z}_p^n$, the value of P at one representative element of L (say the first $x \in L \setminus \{0\}$ according to the coordinate wise order induced by an identification of $\mathbb{Z}_p$ with the set $\{0, \dots, p-1\}$). If we ever need to query the value of P at $x \neq 0$ we can determine it by querying the value of P at the representative of L_x and from this value compute $P(x)$ as if P was linear over L. This way, we are certain that $P(tx) = tP(x)$ for all $x \in \mathbb{Z}_p^n$ and $t \in \mathbb{Z}_p$. Equivalently, we can assume that the oracle function P has this property. We henceforth adopt this convention. Note in particular that this implies that $P(0) = 0$. Taking all the previously introduced conventions into account we perform the following:

Prime Field Linearity Test(P)
1. Randomly choose $x, y, z \in \mathbb{Z}_p^n$ such that $x+y+z = 0$.
2. Reject if $P(x) + P(y) + P(z) \neq 0$.

Henceforth, let $\mathcal{T}$ denote the previous test and $\mathbb{Z}_p^*$ the set $\mathbb{Z}_p \setminus \{0\}$. Also, let ω denote a p–th root of unity. Observe that for $\phi \in \mathbb{Z}_p$, $\phi = 0$ if and only if $\left(\sum_{t \in \mathbb{Z}_p} \omega^{t\phi} \right) / |\mathbb{Z}_p| = 1$. Moreover, $\phi \neq 0$ if and only if $\left(\sum_{t \in \mathbb{Z}_p} \omega^{t\phi} \right) / |\mathbb{Z}_p| = 0$. Hence, for $l \in \mathcal{L}$,

$$\mathsf{Rej}(P, \mathcal{T}) = 1 - \frac{1}{|\mathbb{Z}_p|^{2n}} \sum_{\substack{x,y,z \in \mathbb{Z}_p^n, \\ x+y+z=0}} \left(\frac{1}{|\mathbb{Z}_p|} \sum_{t \in \mathbb{Z}_p} \omega^{t(P(x)+P(y)+P(z))} \right). \quad (4)$$

Now, for two $\mathbb{Z}_p$ valued functions f and g over $\mathbb{Z}_p^n$ denote by ω^f the function that evaluates to $\omega^{f(x)}$ at x and define

$$\chi_g(f) = \frac{1}{|\mathbb{Z}_p|^n} \sum_{x \in \mathbb{Z}_p^n} \omega^{f(x) - g(x)}.$$

Observe that

$$\mathsf{Dist}(f, g) = 1 - \frac{1}{|\mathbb{Z}_p|^n} \sum_{x \in \mathbb{Z}_p^n} \left(\frac{1}{|\mathbb{Z}_p|} \sum_{t \in \mathbb{Z}_p} \omega^{t(f(x) - g(x))} \right) = 1 - \frac{1}{|\mathbb{Z}_p|} \sum_{t \in \mathbb{Z}_p} \chi_{tg}(tf). \quad (5)$$

Lemma 15. *For all $P, l : \mathbb{Z}_p^n \to \mathbb{Z}_p$ such that l is linear and $P(tx) = tP(x)$ for all $x \in \mathbb{Z}_p^n$, $t \in \mathbb{Z}_p$,*

$$\mathsf{Dist}(P, l) = \frac{|\mathbb{Z}_p^*|}{|\mathbb{Z}_p|} (1 - \chi_l(P)).$$

Proof. Note that $tP(x) = P(tx)$ and $tl(x) = l(tx)$ for all $t \in \mathbb{Z}_p$ and $x \in \mathbb{Z}_p^n$. Hence, since multiplication by $t \in \mathbb{Z}_p^*$ induces a permutation of $\mathbb{Z}_p^n$, one has that $\chi_{tl}(tP) = \chi_l(P)$ for all $t \in \mathbb{Z}_p^*$. The conclusion follows from (5) and noting that $\chi_0(\cdot) = 1$.

The following result establishes the (η, η)–robustness of the **Prime Field Linearity Test**.

Lemma 16. *Let $\mathcal{L}$ be the set of linear functions from $\mathbb{Z}_p^n$ to $\mathbb{Z}_p$ and let $P : \mathbb{Z}_p^n \to \mathbb{Z}_p$ be such that $P(tx) = tP(x)$ for all $x \in \mathbb{Z}_p^n$, $t \in \mathbb{Z}_p$. Then,*

$$\mathsf{Rej}(P, \mathcal{T}) = \frac{|\mathbb{Z}_p^*|}{|\mathbb{Z}_p|} \left(1 - \sum_{l \in \mathcal{L}} \left(1 - \frac{|\mathbb{Z}_p|}{|\mathbb{Z}_p^*|} \mathsf{Dist}(P, l) \right)^3 \right) \geq \mathsf{Dist}(P, \mathcal{L}).$$

Proof. Note that $\omega^P = \sum_{l \in \mathcal{L}} \chi_l(P) \omega^l$ and that $t(P(x) + P(y) + P(z)) = P(tx) + P(ty) + P(tz)$. Hence,

$$\omega^{t(P(x) + P(y) + P(z))} = \sum_{l,l',l'' \in \mathcal{L}} \chi_l(P) \chi_{l'}(P) \chi_{l''}(P) \omega^{l(tx) + l'(ty) + l''(tz)}.$$

Furthermore, $z = -(x + y)$ so the linearity of l'' implies that $l''(tz) = -(l''(tx) + l''(ty))$. Thus, (4) yields that

$$\mathsf{Rej}(P, \mathcal{T}) = \frac{|\mathbb{Z}_p^*|}{|\mathbb{Z}_p|} - \sum_{l,l',l'' \in \mathcal{L}} \chi_l(P) \chi_{l'}(P) \chi_{l''}(P) \left(\frac{1}{|\mathbb{Z}_p|} \sum_{t \in \mathbb{Z}_p^*} \chi_{tl''}(tl) \chi_{tl''}(tl') \right).$$

Moreover, when $t \in \mathbb{Z}_p^*$, $\chi_{tl''}(tl)$ and $\chi_{tl''}(tl')$ equal 1 provided $l = l' = l''$, and $\chi_{tl''}(tl)$ or $\chi_{tl''}(tl')$ equal 0 otherwise. Hence,

$$\mathsf{Rej}(P, \mathcal{T}) = \frac{|\mathbb{Z}_p^*|}{|\mathbb{Z}_p|} \left(1 - \sum_{l \in \mathcal{L}} (\chi_l(P))^3 \right).$$

Since $\mathsf{Dist}(P, l)$ is a real number, Lemma 15 implies that so is $\chi_l(P)$. Thus, since $\omega^P = \sum_{l \in \mathcal{L}} \chi_l(P) \omega^l$ and $l(0) = 0$ for every $l \in \mathcal{L}$, we know that $1 = \omega^{P(0)} = \sum_{l \in \mathcal{L}} \chi_l(P)$. Therefore, there is some $l \in \mathcal{L}$ for which $\chi_l(P)$ is non–negative. It follows that,

$$\mathsf{Rej}(P, \mathcal{T}) = \frac{|\mathbb{Z}_p^*|}{|\mathbb{Z}_p|} \left(1 - \sum_{l \in \mathcal{L}} (\chi_l(P))^3 \right) \geq \frac{|\mathbb{Z}_p^*|}{|\mathbb{Z}_p|} \left(1 - \mathop{\mathrm{Max}}_{l \in \mathcal{L}} \chi_l(P) \sum_{l \in \mathcal{L}} (\chi_l(P))^2 \right).$$

By Lemma 15, $\mathrm{Max}_{l \in \mathcal{L}} \chi_l(P) = 1 - (|\mathbb{Z}_p|/|\mathbb{Z}_p^*|) \mathsf{Dist}(P, \mathcal{L})$. The desired conclusion follows by observing that $\sum_{l \in \mathcal{L}} (\chi_l(P))^2 = 1$.

Clearly, the **Prime Field Linearity Test** never rejects a linear function. As far as continuity goes, this is all that usually matters in the PCP context. Note how the verification procedure is simplified both by choosing a prime field structure in which to carry out the work and by forcing structure on the oracle function P, specifically imposing that $P(tx) = tP(x)$ for all $x \in \mathbb{Z}_p^n$ and $t \in \mathbb{Z}_p$. Observe also that the (η, η)–robustness of the test is guaranteed whatever the value of η. Other robustness results discussed in other sections of this work do not exhibit this characteristic. In fact, they typically mean something non–obvious only when η is small. In the PCP context one prefers test analyses that establish that the probability of rejection increases as the distance between the oracle function and the family of functions of interest grows. The majority and median arguments fail to achieve these type of results. The technique on which the proof of Lemma 16 relies was introduced in [BCH+95] and is based on discrete Fourier analysis. This proof technique, in contrast to the majority and median arguments, does not construct a function which is both close to the oracle function and satisfies the property of interest. Hence, when applying the discrete Fourier analysis technique one does not need to assume that the rejection probability of the test is small as is always the case when applying the majority and median argument.

7.2   Property Testing

In the context of testing algebraic functions one is mainly concerned with the problem of determining whether some function to which one has oracle access belongs to some specific class. In the context of property testing one focuses in the case where one has some kind of oracle access to an object, not necessarily a function. Informally, there is an object of which one can ask questions about. The goal is to infer whether or not the object has a specific property. For concreteness, lets consider the following example given by Goldreich [Gol00]: there is a book of which one knows it contains n words and one is allowed to query what its i-th word is — the goal is to determine whether the book is writing in a specific language, say Spanish. As is often the case when testing algebraic functions, if one wants to be completely certain that the book is writing in a specific language one has to query every word. In property testing, as in self–testing, one relaxes the certainty requirement and simply tries to determine whether the object is close or far away from having the property of interest. The notion of distance depends on the problem, e.g., in Goldreich's example, a reasonable choice would be the fraction of non–Spanish words. Thus, suppose that upon seeing one randomly chosen word of the book one decides whether it is writing in Spanish depending on whether the chosen word is a word in such language. Then, a book fully written in Spanish will always be accepted and those books that are at distance δ from being fully written in that language will be discarded with probability δ.

In summary, in property testing one is interested in deciding whether an object has a global property by performing random local checks. One is satisfied if one can distinguish with sufficient confidence between those objects that are close from those that are far from having the global property. In this sense, property testing is a notion of approximation for the aforementioned decision problem.

There are several motivations for the property testing paradigm. When the oracle objects are too large to examine (e.g., the table of a boolean function on a large number of variables) there is no other feasible alternative for deciding whether the object exhibits a given property. Even if the object's size is not large it might be that deciding whether it satisfies the global property is computationally infeasible. In this latter case, property testing provides a reasonable alternative for handling the problem. Finally, when both the oracle object is not too large and the global property can be decided efficiently, property testing might still yield a much faster way of making the correct decision with a high degree of confidence. Moreover, many of the property testers that have been built also allow, at the cost of some additional computational effort, to construct a witness showing the relevant object has the property of interest. These testers could be used to detect instances which are far away from having the property of interest. More expensive computational procedures can thus be run only on instances that have a better chance of having the desired property.

Certainly, exact testing as described earlier in this work can be cast as a property testing problem. Thus, it could be that property testing is a more general paradigm. This is not the case, the two models are mathematically equivalent. One can view property testing as a case of classical testing. However, there are advantages of not doing so, and in recent years the general trend has been to cast new results in the property testing scenario. Indeed, we could have written this whole survey that way. The reasons for not doing so are twofold. The first one is historical: most of the results about algebraic testing were stated in the self–testing context. The second one is specific to this survey. By distinguishing between self–testers, which are algorithms, and (property) tests, which are mathematical objects, we hope that we did clearly point out the difference between the computational and the purely mathematical aspects of the theory. We think that this difference was not adequately dealt with in the previous literature. Had we spoken about property testers and property tests, the difference could have been easily lost for the reader because of the similarity of the terms.

Goldreich, Goldwasser, and Ron [GGR96] were the first to advocate to use of the property testing scenario. In particular they considered the case of testing graph properties. Here, the oracle objects are graphs over a known node set. In [GGR96] the notion of distance between two n–vertex graphs with equal node set is the fraction of edges on which the graphs disagree over n^2. Among the properties considered in [GGR96] were: whether the graph was k–colorable, had a clique containing a ρ fraction of its nodes, had an (edge) cut of size at least ρ fraction of the edges of the complete graph in the same node set, etc. The distance between a graph property is defined in the obvious way, i.e., as the smallest distance between the graph and any graph over the same node set that satisfies the property. In [GR97] a notion of distance better suited to the study of properties of bounded degree graphs was proposed. Specifically, the proposed notion of distance between two n–vertex maximum degree d graphs with equal node set is the fraction of edges on which the graphs disagree over dn. Among the properties studied in [GR97] were: whether the graph was connected, k–vertex–connected, k–edge–connected, planar, etc. Other recent developments in testing graph properties can be found in [GR98,AFKS99,PR99,BR00,GR00].

The works of Goldreich, Goldwasser, and Ron [GGR96,GR97] were influential in shifting the focus from testing algebraic properties of functions to testing non–algebraic properties of different type of objects. Indeed, among other properties/objects that have received attention are: monotonicity of functions [GGLR98,DGL+99], properties of formal languages [AKNS99,New00], geometric properties like clustering [ADPR00,MOP00], and specific properties of quantum gates in quantum circuits [DMMS00].

For surveys on property testing see Goldreich [Gol98] and Ron [Ron00].

References

[ABCG93] S. Ar, M. Blum, B. Codenotti, and P. Gemmell. Checking approximate computations over the reals. In *Proceedings of the 25th Annual ACM Symposium on Theory of Computing*, pages 786–795, San Diego, California, May 1993. ACM.

[ADPR00] N. Alon, S. Dar, M. Parnas, and D. Ron. Testing clustering. In *Proceedings of the 41st Annual Symposium on Foundations of Computer Science*. IEEE, 2000. (To appear).

[AFKS99] N. Alon, E. Fischer, M. Krivelevich, and M. Szegedy. Efficient testing of large graphs. In *Proceedings of the 40th Annual Symposium on Foundations of Computer Science*, pages 656–666, New York City, New York, October 1999. IEEE.

[AKNS99] N. Alon, M. Krivelevich, I. Newman, and M. Szegedy. Regular languages are testable with a constant number of queries. In *Proceedings of the 40th Annual Symposium on Foundations of Computer Science*, pages 645–655, New York City, New York, October 1999. IEEE.

[ALM+92] S. Arora, C. Lund, R. Motwani, M. Sudan, and M. Szegedy. Proof verification and intractability of approximation problems. In *Proceedings of the 33rd Annual Symposium on Foundations of Computer Science*, pages 14–23, Pittsburgh, Pennsylvania, October 1992. IEEE. Final version in [ALM+98].

[ALM+98] S. Arora, C. Lund, R. Motwani, M. Sudan, and M. Szegedy. Proof verification and intractability of approximation problems. *J. of the Association for Computing Machinery*, 45(3):505–555, 1998.

[AS92a] N. Alon and J. H. Spencer. *The probabilistic method*. Wiley–Interscience Series in Discrete Mathematics and Optimization. John Wiley & Sons, Inc., first edition, 1992.

[AS92b] S. Arora and S. Safra. Probabilistic checking of proofs: A new characterization of NP. In *Proceedings of the 33rd Annual Symposium on Foundations of Computer Science*, pages 2–13, Pittsburgh, Pennsylvania, October 1992. IEEE.

[AS97] S. Arora and M. Sudan. Improved low–degree testing and its applications. In *Proceedings of the 29th Annual ACM Symposium on Theory of Computing*, pages 485–495, El Paso, Texas, May 1997. ACM.

[BCH+95] M. Bellare, D. Coppersmith, J. Håstad, M. Kiwi, and M. Sudan. Linearity testing in characteristic two. In *Proceedings of the 36th Annual Symposium on Foundations of Computer Science*, pages 432–441, Milwaukee, Wisconsin, October 1995. IEEE.

[BFL90] L. Babai, L. Fortnow, and C. Lund. Non–deterministic exponential time has two–prover interactive protocols. In *Proceedings of the 31st Annual Symposium on Foundations of Computer Science*, pages 16–25, St. Louis, Missouri, October 1990. IEEE. Final version in [BFL91].

[BFL91] L. Babai, L. Fortnow, and C. Lund. Non–deterministic exponential time has two–prover interactive protocols. *Computational Complexity*, 1:3–40, 1991.

[BFLS91]   L. Babai, L. Fortnow, L. A. Levin, and M. Szegedy. Checking computations in polylogarithmic time. In *Proceedings of the 23rd Annual ACM Symposium on Theory of Computing*, pages 21–31, New Orleans, Louisiana, May 1991. ACM.

[BGLR93]   M. Bellare, S. Goldwasser, C. Lund, and A. Russell. Efficient probabilistically checkable proofs and applications to approximation. In *Proceedings of the 25th Annual ACM Symposium on Theory of Computing*, pages 294–304, San Diego, California, May 1993. ACM.

[BK89]   M. Blum and S. Kannan. Designing programs that check their work. In *Proceedings of the 21st Annual ACM Symposium on Theory of Computing*, pages 86–97, Seattle, Washington, May 1989. ACM. Final version in [BK95].

[BK95]   M. Blum and S. Kannan. Designing programs that check their work. *J. of the Association for Computing Machinery*, 42(1):269–291, 1995.

[BLR90]   M. Blum, M. Luby, and R. Rubinfeld. Self–testing/correcting with applications to numerical problems. In *Proceedings of the 22nd Annual ACM Symposium on Theory of Computing*, pages 73–83, Baltimore, Maryland, May 1990. ACM. Final version in [BLR93].

[BLR93]   M. Blum, M. Luby, and R. Rubinfeld. Self–testing/correcting with applications to numerical problems. *J. of Computer and System Sciences*, 47(3):549–595, 1993.

[Blu88]   M. Blum. Designing programs to check their work. Technical Report TR-88-009, International Computer Science Institure, 1988.

[BR00]   M. Bender and D. Ron. Testing acyclicity of directed graphs in sublinear time. In *Proceedings of the 27th International Colloquium on Automata, Languages and Programming*, volume 1853 of *LNCS*, pages 809–820. Springer–Verlag, 2000.

[BS94]   M. Bellare and M. Sudan. Improved non–approximability results. In *Proceedings of the 26th Annual ACM Symposium on Theory of Computing*, pages 184–193, Montréal, Québec, Canada, May 1994. ACM.

[BW97]   M. Blum and H. Wasserman. Reflections on the Pentium division bug. *IEEE Trans. Comp.*, 26(5):1411–1473, April 1997.

[Cop89]   D. Coppersmith. Manuscript. Result described in [BLR90], December 1989.

[DGL+99]   Y. Dodis, O. Goldreich, E. Lehman, S. Rsakhodnikova, D. Ron, and A. Samorodnitsky. Improved testing algorithms for monotonicity. In *Proceedings of RANDOM'99*, volume 1671 of *LNCS*, pages 97–108. Springer–Verlag, 1999.

[DMMS00]   W. van Dam, F. Magniez, M. Mosca, and M. Santha. Self–testing of universal and fault–tolerant sets of quantum gates. In *Proceedings of the 32nd Annual ACM Symposium on Theory of Computing*, pages 688–696, Portland, Oregon, May 2000. ACM.

[EKR96]   F. Ergün, S. Ravi Kumar, and R. Rubinfeld. Approximate checking of polynomials and functional equations. In *Proceedings of the 37th Annual Symposium on Foundations of Computer Science*, pages 592–601, Burlington, Vermont, October 1996. IEEE.

[Erg95]   F. Ergün. Testing multivariate linear functions: Overcoming the generator bottleneck. In *Proceedings of the 27th Annual ACM Symposium on Theory of Computing*, pages 407–416, Las Vegas, Nevada, May 1995. ACM.

[ESK00]   F. Ergün, S. Sivakumar, and S. Ravi Kumar. Self–testing without the generator bottleneck. *SIAM J. on Computing*, 29(5):1630–1651, 2000.

[FFT]   FFTW is a free collection of fast C routines for computing the Di-
            screte Fourier Transform in one or more dimensions. For more details see
            www.fftw.org.

[FGL+91]   U. Feige, S. Goldwasser, L. Lovász, S. Safra, and M. Szegedy. Approxi-
            mating clique is almost NP–complete. In *Proceedings of the 32nd Annual
            Symposium on Foundations of Computer Science*, pages 2–12, San Juan,
            Puerto Rico, October 1991. IEEE.

[For95]   G. L. Forti. Hyers–Ulam stability of functional equations in several varia-
            bles. *Aequationes Mathematicae*, 50:143–190, 1995.

[GGLR98]   O. Goldreich, S. Goldwasser, E. Lehman, and D. Ron. Testing monoto-
            nicity. In *Proceedings of the 39th Annual Symposium on Foundations of
            Computer Science*, pages 426–435, Palo Alto, California, November 1998.
            IEEE.

[GGR96]   O. Goldreich, S. Goldwasser, and D. Ron. Property testing and its connec-
            tion to learning and approximation. In *Proceedings of the 37th Annual
            Symposium on Foundations of Computer Science*, pages 339–348, Burling-
            ton, Vermont, October 1996. IEEE.

[GLR+91]   P. Gemmell, R. Lipton, R. Rubinfeld, M. Sudan, and A. Wigderson. Self–
            testing/correcting for polynomials and for approximate functions. In *Pro-
            ceedings of the 23rd Annual ACM Symposium on Theory of Computing*,
            pages 32–42, New Orleans, Louisiana, May 1991. ACM.

[Gol98]   O. Goldreich. *Combinatorial property testing — A survey*, volume 43 of *DI-
            MACS Series in Discrete Mathematics and Theoretical Computer Science*,
            pages 45–60. ACM/AMS, 1998.

[Gol00]   O. Goldreich. Talk given at the DIMACS Workshop on Sublinear Algo-
            rithms, September 2000.

[GR97]   O. Goldreich and D. Ron. Property testing in bounded degree graphs. In
            Proceedings of the 29th Annual ACM Symposium on Theory of Computing,
            pages 406–415, El Paso, Texas, May 1997. ACM.

[GR98]   O. Goldreich and D. Ron. A sublinear bipartiteness tester for bounded
            degree graphs. In *Proceedings of the 30th Annual ACM Symposium on
            Theory of Computing*, pages 289–298, Dallas, Texas, May 1998. ACM.

[GR00]   O. Goldreich and D. Ron. On testing expansion in bounded–degree graphs.
            Technical Report ECCC TR00–020, Electronic Colloquium on Computa-
            tional Complexity, 2000. (Available at www.eccc.uni-trier.de/eccc/).

[Hås96]   J. Håstad. Testing of the long code and hardness of clique. In *Proceedings of
            the 37nd Annual IEEE Symposium on Foundations of Computer Science*,
            pages 11–19, Burlington, Vermont, October 1996. IEEE.

[Hås97]   J. Håstad. Getting optimal in–approximability results. In *Proceedings of
            the 31st Annual ACM Symposium on Theory of Computing*, pages 1–10,
            El Paso, Texas, May 1997. ACM.

[HR92]   D. H. Hyers and T. M. Rassias. Approximate homomorphisms. *Aequatio-
            nes Mathematicae*, 44:125–153, 1992.

[Hye41]   D. H. Hyers. On the stability of the linear functional equation. *Proceedings
            of the National Academy of Science, U.S.A.*, 27:222–224, 1941.

[Kiw96]   M. Kiwi. *Probabilistically Checkable Proofs and the Testing of Hadamard–
            like Codes*. PhD thesis, Massachusetts Institute of Technology, February
            1996.

[KMS99]   M. Kiwi, F. Magniez, and M. Santha. Approximate testing with relative
            error. In *Proceedings of the 31st Annual ACM Symposium on Theory of
            Computing*, pages 51–60, Atlanta, Georgia, May 1999. ACM.

[KR97]   M. Kiwi and A. Russell. Linearity testing over prime fields. Unpublished manuscript, 1997.

[Lip91]   R. J. Lipton. *New directions in testing*, volume 2 of *DIMACS Series in Discrete Mathematics and Theoretical Computer Science*, pages 191–202. ACM/AMS, 1991.

[Mag00a]   F. Magniez. *Auto–test pour les calculs approché et quantique*. PhD thesis, Université Paris–Sud, France, 2000.

[Mag00b]   F. Magniez. Multi–linearity self–testing with relative error. In *Proceedings of the 17th Annual Symposium on Theoretical Aspects of Computer Science*, volume 1770 of *LNCS*, pages 302–313. Springer–Verlag, 2000.

[MOP00]   M. Mishra, D. Oblinger, and L. Pirtt. Way–sublinear time approximate (PAC) clustering. Unpublished, 2000.

[New00]   I. Newman. Testing of functions that have small width branching programs. In *Proceedings of the 41st Annual Symposium on Foundations of Computer Science*. IEEE, 2000. (To appear).

[PR99]   M. Parnas and D. Ron. Testing the diameter of graphs. In *Proceedings of RANDOM'99*, volume 1671 of *LNCS*, pages 85–96. Springer–Verlag, 1999.

[PS94]   A. Polishchuk and D. Spielman. Nearly–linear size holographic proofs. In *Proceedings of the 26th Annual ACM Symposium on Theory of Computing*, pages 194–203, Montréal, Québec, Canada, May 1994. ACM.

[Ron00]   D. Ron. Property testing (A tutorial), 2000. (Available at www.eng.tau.ac.il/~danar/papers.html). To appear in *Handbook on Randomization*.

[Rub90]   R. Rubinfeld. *A mathematical theory of self–checking, self–testing and self–correcting programs*. PhD thesis, University of California, Berkeley, 1990.

[Rub94]   R. Rubinfeld. On the robustness of functional equations. In *Proceedings of the 35th Annual Symposium on Foundations of Computer Science*, pages 288–299, Santa Fe, New Mexico, November 1994. IEEE. Final version in [Rub99].

[Rub99]   R. Rubinfeld. On the robustness of functional equations. *SIAM J. on Computing*, 28(6):1972–1997, 1999.

[RS92a]   T. M. Rassias and P. Šemrl. On the behaviour of mappings which do not satisfy Hyers–Ulam stability. *Proceedings of the American Mathematical Society*, 114(4):989–993, April 1992.

[RS92b]   R. Rubinfeld and M. Sudan. Testing polynomial functions efficiently and over rational domains. In *Proceedings of the 3rd Annual ACM–SIAM Symposium on Discrete Algorithms*, pages 23–32, Orlando, Florida, January 1992. ACM/SIAM. Final version in [RS96].

[RS96]   R. Rubinfeld and M. Sudan. Robust characterizations of polynomials with applications to program testing. *SIAM Journal of Computing*, 25(2):252–271, April 1996.

[Sko83]   F. Skopf. Sull'approssimazione delle applicazioni localmente δ–additive. *Atti della Accademia delle Sciencze di Torino*, 117:377–389, 1983. (In Italian.).

[Tre98]   L. Trevisan. Recycling queries in PCPs and in linearity tests. In *Proceedings of the 30th Annual ACM Symposium on Theory of Computing*, pages 299–308, Dallas, Texas, May 1998. ACM.

The Regularity Lemma and Its Applications in Graph Theory[*]

János Komlós[1,2], Ali Shokoufandeh[3], Miklós Simonovits[1,2], and
Endre Szemerédi[1,2]

[1] Rutgers University
New Brunswick, NJ
[2] Hungarian Academy of Sciences
[3] Department of Mathematics and Computer Science
Drexel University, Philadelphia, PA

Abstract. Szemerédi's Regularity Lemma is an important tool in discrete mathematics. It says that, in some sense, all graphs can be approximated by random-looking graphs. Therefore the lemma helps in proving theorems for arbitrary graphs whenever the corresponding result is easy for random graphs. In the last few years more and more new results were obtained by using the Regularity Lemma, and also some new variants and generalizations appeared. Komlós and Simonovits have written a survey on the topic [96]. The present survey is, in a sense, a continuation of the earlier survey. Here we describe some sample applications and generalizations. To keep the paper self-contained we decided to repeat (sometimes in a shortened form) parts of the first survey, but the emphasis is on new results.

Preface

The Regularity Lemma [127] is one of the most powerful tools of (extremal) graph theory. It was invented as an auxiliary lemma in the proof of the famous conjecture of Erdős and Turán [53] that sequences of integers of positive upper density must always contain long arithmetic progressions. Its basic content could be described by saying that every graph can, in some sense, be well approximated by random graphs. Since random graphs of a given edge density are much easier to treat than all graphs of the same edge-density, the Regularity Lemma helps us to carry over results that are trivial for random graphs to the class of all graphs with a given number of edges. It is particularly helpful in "fuzzy" situations, i.e., when the conjectured extremal graphs have no transparent structure.

Remark. Sometimes the Regularity Lemma is called Uniformity Lemma, see e.g., [61] and [4].

Notation. In this paper we only consider simple graphs – undirected graphs without loops and multiple edges: $G = (V, E)$ where $V = V(G)$ is the vertex-set

---

[*] Research supported in part by the NSF grant DMS-9801396.

G.B. Khosrovshahi et al. (Eds.): Theoretical Aspects of Computer Science, LNCS 2292, pp. 84–112, 2002.

of G and $E = E(G) \subset \binom{V}{2}$ is the edge-set of G. $v(G) = |V(G)|$ is the number of vertices in G (order), $e(G) = |E(G)|$ is the number of edges in G (size). G_n will always denote a graph with n vertices. $deg(v)$ is the degree of vertex v and $deg(v, Y)$ is the number of neighbours of v in Y. $\delta(G), \Delta(G)$ and $t(G)$ are the minimum degree, maximum degree and average degree of G. $\chi(G)$ is the chromatic number of G. $N(x)$ is the set of neighbours of the vertex x, and $e(X, Y)$ is the number of edges between X and Y. A bipartite graph G with color-classes A and B and edge-set E will sometimes be written as $G = (A, B, E)$, $E \subset A \times B$. For disjoint X, Y, we define the **density**

$$d(X, Y) = \frac{e(X, Y)}{|X| \cdot |Y|}.$$

$G(U)$ is the restriction of G to U and $G - U$ is the restriction of G to $V(G) - U$. For two disjoint subsets A, B of $V(G)$, we write $G(A, B)$ for the subgraph with vertex set $A \cup B$ whose edges are those of G with one endpoint in A and the other in B.

For graphs G and H, $H \subset G$ means that H is a subgraph of G, but often we will use this in the looser sense that G has a subgraph isomorphic to H (H is embeddable into G), that is, there is a one-to-one map (injection) $\varphi : V(H) \to V(G)$ such that $\{x, y\} \in E(H)$ implies $\{\varphi(x), \varphi(y)\} \in E(G)$. $\|H \to G\|$ denotes the number of labelled copies of H in G. The cardinality of a set S will mostly be denoted by $|S|$, but sometimes we write $\#S$. We will be somewhat sloppy by often disregarding rounding.

1   Introduction

1.1   The Structure of This Survey

We will start with some historical remarks, then we state the Regularity Lemma. After that we introduce the basic notion of the Reduced Graph of a graph corresponding to a partition of the vertex-set, and state a simple but useful tool (Embedding Lemma). The much stronger version called Blow-up Lemma is mentioned later. The latter has found many applications since [96] was published. (For a short survey on the Blow-up Lemma, see [87].)

We will also touch upon some algorithmic aspects of the Regularity Lemma, its relation to quasi-random graphs and extremal subgraphs of a random graph. We also shortly mention a sparse version.

The results quoted here only serve as illustrations; we did not attempt to write a comprehensive survey. An extended version is planned in the near future.

1.2   Regular Pairs

Regular pairs are highly uniform bipartite graphs, namely ones in which the density of any reasonably sized subgraph is about the same as the overall density of the graph.

Definition 1 (Regularity condition). *Let $\varepsilon > 0$. Given a graph G and two disjoint vertex sets $A \subset V$, $B \subset V$, we say that the pair (A, B) is ε-regular if for every $X \subset A$ and $Y \subset B$ satisfying*

$$|X| > \varepsilon|A| \quad \text{and} \quad |Y| > \varepsilon|B|$$

we have

$$|d(X, Y) - d(A, B)| < \varepsilon.$$

The next one is the most important property of regular pairs.

Fact 1 (Most degrees into a large set are large) *Let (A, B) be an ε-regular pair with density d. Then for any $Y \subset B$, $|Y| > \varepsilon|B|$ we have*

$$\#\{x \in A : deg(x, Y) \le (d - \varepsilon)|Y|\} \le \varepsilon|A|.$$

For other basic properties of regular pairs see [96].
We will also use another version of regularity:

Definition 2 (Super-regularity). *Given a graph G and two disjoint vertex sets $A \subset V$, $B \subset V$, we say that the pair (A, B) is (ε, δ)-super-regular if for every $X \subset A$ and $Y \subset B$ satisfying*

$$|X| > \varepsilon|A| \quad \text{and} \quad |Y| > \varepsilon|B|$$

we have

$$e(X, Y) > \delta|X||Y|,$$

and furthermore,

$$deg(a) > \delta|B| \quad \text{for all} \quad a \in A, \quad \text{and} \quad deg(b) > \delta|A| \quad \text{for all} \quad b \in B.$$

1.3   The Regularity Lemma

The Regularity Lemma says that every dense graph can be partitioned into a small number of regular pairs and a few leftover edges. Since regular pairs behave as random bipartite graphs in many ways, the Regularity Lemma provides us with an approximation of a large dense graph with the union of a small number of random-looking bipartite graphs.

Theorem 2 (Regularity Lemma, Szemerédi 1978 [127]). *For every $\varepsilon > 0$ there exists an integer $M = M(\varepsilon)$ with the following property: for every graph G there is a partition of the vertex set into k classes $V = V_1 + V_2 + \ldots + V_k$ such that*

- *$k \le M$,*
- *$|V_i| \le \lceil \varepsilon|V| \rceil$ for every i,*
- *$||V_i| - |V_j|| \le 1$ for all i, j (equipartition),*
- *(V_i, V_j) is ε-regular in G for all but at most εk^2 pairs (i, j).*

The classes V_i will be called **groups** or **clusters**.

If we delete the edges within clusters as well as edges that belong to irregular pairs of the partition, we get a subgraph $G' \subset G$ that is more uniform, more random-looking, and therefore more manageable. Since the number of edges deleted is small compared to $|V|^2$, the Regularity Lemma provides us with a good approximation of G by the random-looking graph G'. Of course, if we have a sequence (G_n) of graphs with $e(G_n) = o(n^2)$, the Regularity Lemma becomes trivial: G_n are approximated by empty graphs. Thus the Regularity Lemma is useful only for **large, dense graphs.**

Remark 1. A drawback of the result is that the bound obtained for $M(\varepsilon)$ is extremely large, namely a tower of 2's of height proportional to ε^{-5}. That this is not a weakness of Szemerédi's proof but rather an inherent feature of the Regularity Lemma was shown by Timothy Gowers [70] (see also [9]).

The Regularity Lemma asserts in a way that every graph can be approximated by generalized random graphs.

Definition 3 ([118]). *Given an $r \times r$ symmetric matrix (p_{ij}) with $0 \le p_{ij} \le 1$, and positive integers $n_1, \ldots, n_r$, we define a **generalized random graph** R_n (for $n = n_1 + \cdots + n_r$) by partitioning n vertices into classes V_i of size n_i and then joining the vertices $x \in V_i$, $y \in V_j$ with probability p_{ij}, independently for all pairs $\{x, y\}$.*

Remark 2. Often, the application of the Regularity Lemma makes things transparent but the same results can be achieved without it equally easily. One would like to know when one can replace the Regularity Lemma with "more elementary" tools and when the application of the Regularity Lemma is unavoidable. The basic experience is that when in the conjectured extremal graphs for a problem the densities in the regular partition are all near to 0 or 1, then the Regularity Lemma can probably be eliminated. On the other hand, if these densities are strictly bounded away from 0 and 1 then the application of the Regularity Lemma is often unavoidable.

1.4   The Road to the Regularity Lemma

The following is a basic result in combinatorial number theory.

Theorem 3 (van der Waerden 1927 [131]). *Let k and t be arbitrary positive integers. If we color the integers with t colors, at least one color-class will contain an arithmetic progression of k terms.*

A standard compactness argument shows that the following is an equivalent form.

Theorem 4 (van der Waerden - finite version). *For any integers k and t there exists an n such that if we color the integers $\{1, \ldots, n\}$ with t colors, then at least one color-class will contain an arithmetic progression of k terms.*

This is a Ramsey type theorem in that it only claims the existence of a given configuration in one of the color classes without getting any control over which class it is. It turns out that the van der Waerden problem is *not* a true Ramsey type question but of a density type: the only thing that matters is that at least one of the color classes contains relatively many elements. Indeed, answering a very deep and difficult conjecture of P. Erdős and P. Turán from 1936 [53], Endre Szemerédi proved that positive upper density implies the existence of an arithmetic progression of k terms.

Theorem 5 (Szemerédi 1975 [126]). *For every integer $k > 2$ and $\varepsilon > 0$ there exists a threshold $n_0 = n_0(k, \varepsilon)$ such that if $n \geq n_0$, $A \subset \{1, \ldots, n\}$ and $|A| > \varepsilon n$, then A contains an arithmetic progression of k terms.*

Remark. For $k = 3$ this is a theorem of K.F. Roth [103] that dates back to 1954, and it was already an important breakthrough when Szemerédi succeeded in proving the theorem in 1969 for $k = 4$ [124]. One of the interesting questions in this field is the speed of convergence to 0 of $r_k(n)/n$, where $r_k(n)$ is the maximum size of a subset of $[n]$ not containing an arithmetic progression of length k. Szemerédi's proof used van der Waerden's theorem and therefore gave no reasonable bound on the convergence rate of $r_4(n)/n$. Roth found an analytical proof a little later [104,105] not using van der Waerden's theorem and thus providing the first meaningful estimates on the convergence rate of $r_4(n)/n$ [104].

Szemerédi's theorem (for general k) was also proved by Fürstenberg [66] in 1977 using ergodic theoretical methods. It was not quite clear first how different the Fürstenberg proof was from that of Szemerédi, but subsequent generalizations due to Fürstenberg and Katznelson [68] and later by Bergelson and Leibman [7] convinced the mathematical community that Ergodic Theory is a natural tool to attack combinatorial questions. The narrow scope of this survey does not allow us to explain these generalizations. We refer the reader to the book of R.L. Graham, B. Rothschild and J. Spencer, *Ramsey Theory* [71], which describes the Hales-Jewett theorem and how these theorems are related, and its chapter "Beyond Combinatorics" gives an introduction into related subfields of topology and ergodic theory. Another good source is the paper of Fürstenberg [67].

2   Early Applications

Among the first graph theoretical applications, the Ramsey-Turán theorem for K_4 and the $(6, 3)$-theorem of Ruzsa and Szemerédi were proved using (an earlier version of) the Regularity Lemma.

2.1   The (6,3)-Problem

The $(6, 3)$-problem is a special hypergraph extremal problem: Brown, Erdős and T. Sós asked for the determination of the maximum number of hyperedges an r-uniform hypergraph can have without containing ℓ hyperedges the union of which is at most k [17,16]. One of the simplest cases they could not settle was this $(6,3)$-problem.

Theorem 6 (The $(6,3)$-theorem, Ruzsa-Szemerédi 1976 [111]). *If H_n is a 3-uniform hypergraph on n vertices not containing 6 points with 3 or more triples, then $e(H_n) = o(n^2)$.*

(Since the function $M(\varepsilon)$ grows incredibly fast, this would only give an upper bound $r_3(n) = O(n/\log^* n)$, much weaker than Roth's $r_3(n) = O(n/\log\log n)$, let alone the often conjectured $r_3(n) = O(n/\log n)$. The best known upper bound is due to Heath-Brown [80] and to Szemerédi [128] improving Heath-Brown's result, according to which $r_3(n) \le O(n/\log^{1/4-\varepsilon} n)$.)

The $(6,3)$ theorem was generalized by Erdős, Frankl and Rödl as follows. Let $g_r(n, v, e)$ denote the maximum number of r-edges an r-uniform hypergraph may have if the union of any e edges span more than v vertices.

Theorem 7 (Erdős-Frankl-Rödl [38]). *For all (fixed) r, $g_r(n, 3r-3, 3) = o(n^2)$.*

For another strengthening of the $(6,3)$ theorem, see [32].

2.2 Applications in Ramsey-Turán Theory

Theorem 8 (Ramsey-Turán for K_4, Szemerédi 1972 [125]). *If G_n contains no K_4 and only contains $o(n)$ independent vertices, then $e(G_n) < \frac{1}{8}n^2 + o(n^2)$.*

Remark. Since most people believed that in Theorem 8 the upper bound $n^2/8$ can be improved to $o(n^2)$, it was quite a surprise when in 1976 Bollobás and Erdős [10] came up with an ingenious geometric construction which showed that the constant $1/8$ in the theorem is best possible. That is, they showed the existence of a graph sequence (H_n) for which

$$K_4 \not\subset H_n, \quad \alpha(H_n) = o(n) \quad \text{and} \quad e(H_n) > \frac{n^2}{8} - o(n^2).$$

Remark. A typical feature of the application of the regularity lemma can be seen above, namely that we do not distinguish between $o(n)$ and $o(m)$, since the number k of clusters is bounded (in terms of ε only) and $m \sim n/k$.

Remark. The problem of determining $\max e(G_n)$ under the condition

$$K_p \not\subset G_n \quad \text{and} \quad \alpha(G_n) = o(n)$$

is much easier for odd p than for even p. A theorem of Erdős and T. Sós [51] describing the odd case was a starting point of the theory of **Ramsey-Turán problems**. The next important contribution was the above-mentioned theorem of Szemerédi (and then the counterpart due to Bollobás and Erdős). Finally the paper of Erdős, Hajnal, T. Sós and Szemerédi [45] completely solved the problem for all even p by generalizing the above Szemerédi-Bollobás-Erdős theorems. It also used the Regularity Lemma.

One reason why the Regularity Lemma can be used here is that if we know that the reduced graph contains some graph L, (e.g., a K_3), then using the $o(n)$-condition we can guarantee a larger subgraph (e.g., a K_4) in the original graph. According to our philosophy, one reason why probably the use of the Regularity Lemma is unavoidable is that the edge-density in the conjectured extremal graph is $1/2$; bounded away from 0 and 1.

There are many related Ramsey-Turán theorems; we refer the reader to [43] and [44], or to the survey [121]. The very first Ramsey-Turán type problem can be found in the paper [122] of Vera T. Sós.

2.3   Building Small Induced Subgraphs

While the reduced graph R of G certainly reflects many aspects of G, when discussing induced subgraphs the definition should be changed in a natural way. Given a partition $V_1, \ldots, V_k$ of the vertex-set V of G and positive parameters ε, d, we define the induced reduced graph as the graph whose vertices are the clusters $V_1, \ldots, V_k$ and V_i and V_j are adjacent if the pair (V_i, V_j) is ε-regular in G with density *between* d and $1 - d$.

Below we will describe an application of the regularity lemma about the existence of small induced subgraphs of a graph, not by assuming that the graph has many edges but by putting some condition on the graph which makes its structure randomlike, fuzzy.

Definition 4. *A graph $G = (V, E)$ has the property (γ, δ, σ) if for every subset $S \subset V$ with $|S| > \gamma|V|$ the induced graph $G(S)$ satisfies*

$$(\sigma - \delta)\binom{|S|}{2} \leq e(G(S)) \leq (\sigma + \delta)\binom{|S|}{2}.$$

Theorem 9 (Rödl 1986 [107]). *For every positive integer k and every $\sigma > 0$ and $\delta > 0$ such that $\delta < \sigma < 1 - \delta$ there exists a γ and a positive integer n_0 such that every graph G_n with $n \geq n_0$ vertices satisfying the property (γ, δ, σ) contains all graphs with k vertices as induced subgraphs.*

Rödl also points out that this theorem yields an easy proof (see [101]) of the following generalization of a Ramsey theorem first proved in [28,42] and [106]:

Theorem 10. *For every graph L there exists a graph H such that for any 2-coloring of the edges of H, H must contain an induced monochromatic L.*

The next theorem of Rödl answers a question of Erdős [8,36].

Theorem 11. *For every positive integer k and positive σ and γ there exists a $\delta > 0$ and a positive integer n_0 such that every graph G_n with at least n_0 vertices having property (γ, δ, σ) contains all graphs with k vertices as induced subgraphs.*

(Erdős asked if the above theorem holds for $\frac{1}{2}, \delta, \frac{1}{2}$ and K_k.)

The reader later may notice the analogy and the connection between this theorem and some results of Chung, Graham and Wilson on quasi-random graphs (see Section 8).

2.4 Diameter-Critical Graphs

We shall need a notation: If H is an arbitrary graph with vertex set $\{x_1, \ldots, x_k\}$ and $a_1, \ldots, a_k$ are non-negative integers, then $H(a_1, \ldots, a_k)$ denotes the graph obtained from H_k by replacing x_j by a set X_i of a_i independent vertices, and joining each $x \in X_i$ to each $x' \in X_j$ for $1 \le i < j \le k$ exactly if $(x_i, x_j) \in E(H)$.

Consider all graphs G_n of diameter 2. The minimum number of edges in such graphs is attained by the star $K(1, n-1)$. There are many results on graphs of diameter 2. An interesting subclass is the class of 2-diameter-critical graphs. These are minimal graphs of diameter 2: deleting any edge we get a graph of diameter greater than 2. The cycle C_5 is one of the simplest 2-diameter-critical graphs. If H is a 2-diameter-critical graph, then $H(a_1, \ldots, a_k)$ is also 2-diameter-critical. So $T_{n,2}$, and more generally of $K(a, b)$, are 2-diameter-critical. Independently, Murty and Simon (see in [21]) formulated the following conjecture:

Conjecture 12 *If G_n is a minimal graph of diameter 2, then $e(G) \le \lfloor n^2/4 \rfloor$. Equality holds if and only if G_n is the complete bipartite graph $K_{\lfloor n/2 \rfloor, \lceil n/2 \rceil}$.*

Füredi used the Regularity Lemma to prove this.

Theorem 13 (Füredi 1992 [65]). *Conjecture 12 is true for $n \ge n_0$.*

Here is an interesting point: Füredi did not need the whole strength of the Regularity Lemma, only a consequence of it, the $(6, 3)$-theorem.

3 How to Apply the Regularity Lemma

3.1 The Reduced Graph

Given an arbitrary graph $G = (V, E)$, a partition P of the vertex-set V into $V_1, \ldots, V_k$, and two parameters ε, d, we define the **Reduced Graph** (or Cluster graph) R as follows: its vertices are the clusters $V_1, \ldots, V_k$ and V_i is joined to V_j if (V_i, V_j) is ε-regular with density more than d. Most applications of the Regularity Lemma use Reduced Graphs, and they depend upon the fact that many properties of R are inherited by G.

The most important property of Reduced Graphs is mentioned in the following section.

3.2 A Useful Lemma

Many of the proofs using the Regularity Lemma struggle through similar technical details. These details are often variants of an essential feature of the Regularity Lemma: If G has a reduced graph R and if the parameter ε is small enough, then every small subgraph H of R is also a subgraph of G. In the first applications of the Regularity Lemma the graph H was fixed, but the greedy algorithm outlined in the section "Building up small subgraphs" works smoothly even when the order of H is proportional with that of G as long as H has bounded degrees. (Another standard class of applications - embedding trees into dense graphs - will be discussed later.)

The above mentioned greedy embedding method for bounded degree graphs is so frequently used that, just to avoid repetitions of technical details, it is worth while spelling it out in a quotable form.

For a graph R and positive integer t, let $R(t)$ be the graph obtained from R by replacing each vertex $x \in V(R)$ by a set V_x of t independent vertices, and joining $u \in V_x$ to $v \in V_y$ iff (x, y) is an edge of R. In other words, we replace the edges of R by copies of the complete bipartite graph $K_{t,t}$.

Theorem 14 (Embedding Lemma). *Given $d > \varepsilon > 0$, a graph R, and a positive integer m, let us construct a graph G by replacing every vertex of R by m vertices, and replacing the edges of R with ε-regular pairs of density at least d. Let H be a subgraph of $R(t)$ with h vertices and maximum degree $\Delta > 0$, and let $\delta = d - \varepsilon$ and $\varepsilon_0 = \delta^\Delta/(2 + \Delta)$. If $\varepsilon \leq \varepsilon_0$ and $t - 1 \leq \varepsilon_0 m$, then $H \subset G$. In fact,*
$$\|H \to G\| > (\varepsilon_0 m)^h.$$

Remark. Note that $v(R)$ didn't play any role here.

Remark. Often we use this for R itself (that is, for $t = 1$): If $\varepsilon \leq \delta^{\Delta(R)}/(2 + \Delta(R))$ then $R \subset G$, in fact, $\|R \to G\| \geq (\varepsilon m)^{v(R)}$.

Remark. Using the fact that large subgraphs of regular pairs are still regular (with a different value of ε), it is easy to replace the condition $H \subset R(\varepsilon_0 m)$ with the assumptions

(*)   every component of H is smaller than $\varepsilon_0 m$,
(**) $H \subset R((1 - \varepsilon_0)m)$.

Most of the classical proofs using the Regularity Lemma can be simplified by the application of the Embedding Lemma. However, this only helps presentability; the original proof ideas – basically building up subgraphs vertex-by-vertex – are simply summarized in the Embedding Lemma.

One can strengthen the lemma tremendously by proving a similar statement for all bounded degree subgraphs H of the full $R(m)$. This provides a very powerful tool (Blow-up Lemma), and it is described in Section 4.6.

Proof of the Embedding Lemma. We prove the following more general estimate.

If $t - 1 \leq (\delta^\Delta - \Delta\varepsilon)m$ then $\|H \to G\| > \left[(\delta^\Delta - \Delta\varepsilon)m - (t - 1)\right]^h$.

We embed the vertices $v_1, \ldots, v_h$ of H into G by picking them one-by-one. For each v_j not picked yet we keep track of an ever shrinking set C_{ij} that v_j is confined to, and we only make a final choice for the location of v_j at time j. At time 0, C_{0j} is the full m-set v_j is *a priori* restricted to in the natural way. Hence $|C_{0j}| = m$ for all j. The algorithm at time $i \geq 1$ consists of two steps.

Step 1 - *Picking v_i.* We pick a vertex $v_i \in C_{i-1,i}$ such that

$$deg_G(v_i, C_{i-1,j}) > \delta |C_{i-1,j}| \quad \text{for all} \quad j > i \quad \text{such that} \quad \{v_i, v_j\} \in E(H). \quad (1)$$

Step 2. - *Updating the C_j's.* We set, for each $j > i$,

$$C_{ij} = \begin{cases} C_{i-1,j} \cap N(v_i) & \text{if } \{v_i, v_j\} \in E(H) \\ C_{i-1,j} & \text{otherwise.} \end{cases}$$

For $i < j$, let $d_{ij} = \#\{\ell \in [i] : \{v_\ell, v_j\} \in E(H)\}$.

Fact.   *If $d_{ij} > 0$ then $|C_{ij}| > \delta^{d_{ij}} m$. (If $d_{ij} = 0$ then $|C_{ij}| = m$.)*

Thus, for all $i < j$, $|C_{ij}| > \delta^\Delta m \geq \varepsilon m$, and hence, when choosing the exact location of v_i, all but at most $\Delta \varepsilon m$ vertices of $C_{i-1,i}$ satisfy (1). Consequently, we have at least

$$|C_{i-1,i}| - \Delta \varepsilon m - (t-1) > (\delta^\Delta - \Delta \varepsilon)m - (t-1)$$

free choices for v_i, proving the claim.   ∎

Remark. We did not use the full strength of ε-regularity for the pairs (A, B) of m-sets replacing the edges of H, only the following one-sided property:

$$X \subset A, |X| > \varepsilon |A|, Y \subset B, |Y| > \varepsilon |B| \quad \text{imply} \quad e(X, Y) > \delta |X||Y|.$$

We already mentioned that in a sense the Regularity Lemma says that all graphs can be approximated by generalized random graphs. The following observation was used in the paper of Simonovits and T. Sós [118] to characterize quasi-random graphs.

Theorem 15. *Let $\delta > 0$ be arbitrary, and let $V_0, V_1, \ldots, V_k$ be a regular partition of an arbitrary graph G_n with $\varepsilon = \delta^2$ and each cluster size less than δn. Let Q_n be the random graph obtained by replacing the edges joining the classes V_i and V_j (for all $i \neq j$) by independently chosen random edges of probability $p_{i,j} := d(V_i, V_j)$, and let H be any graph with ℓ vertices. If $n \geq n_0$, then*

$$\|H \to Q_n\| - C_\ell \delta n^\ell \leq \|H \to G_n\| \leq \|H \to Q_n\| + C_\ell \delta n^\ell.$$

almost surely, where C_ℓ is a constant depending only on ℓ.

Most applications start with applying the Regularity Lemma for a graph G and finding the corresponding Reduced Graph R. Then usually a classical extremal graph theorem (like the König-Hall theorem, Dirac's theorem, Turán's theorem or the Hajnal-Szemerédi theorem) is applied to the graph R. Then an argument similar to the Embedding Lemma (or its strengthened version, the Blow-up Lemma) is used to lift the theorem back to the graph G.

3.3   Some Classical Extremal Graph Theorems

This is only a brief overview of the standard results from extremal graph theory most often used in applications of the Regularity Lemma. For a detailed description of the field we refer the reader to [8,117,64].

The field of extremal graph theory started with the historical paper of Pál Turán in 1941, in which he determined the minimal number of edges that guarantees the existence of a p-clique in a graph. The following form is somewhat weaker than the original theorem of Turán, but it is perhaps the most usable form.

Theorem 16 (Turán 1941 [130]). *If G_n is a graph with n vertices and*

$$e(G) > \left(1 - \frac{1}{p-1}\right)\frac{n^2}{2},$$

then $K_p \subset G_n$.

In general, given a family $\mathcal{L}$ of **excluded** graphs, one would like to find the maximum number of edges a graph G_n can have without containing any subgraph $L \in \mathcal{L}$. This maximum is denoted by $\mathrm{ex}(n, \mathcal{L})$ and the graphs attaining the maximum are called **extremal graphs**. (We will use the notation $\mathrm{ex}(n, \mathcal{L})$ for hypergraphs, too.) These problems are often called Turán type problems, and are mostly considered for **simple graphs or hypergraphs**, but there are also many results for **multigraphs and digraphs** of bounded edge- or arc-multiplicity (see e.g. [13,14,15,18,114]).

Using this notation, the above form of Turán's theorem says that

$$\mathrm{ex}(n, K_p) \leq \left(1 - \frac{1}{p-1}\right)\frac{n^2}{2}.$$

The following theorem of Erdős and Stone determines $\mathrm{ex}(n, K_p(t, \ldots, t))$ asymptotically.

Theorem 17 (Erdős-Stone 1946 [52] - Weak Form). *For any integers $p \geq 2$ and $t \geq 1$,*

$$\mathrm{ex}(n, K_p(t, \ldots, t)) = \left(1 - \frac{1}{p-1}\right)\binom{n}{2} + o(n^2).$$

(For strengthened versions, see [25,26].) This is, however, much more than just another Turán type extremal result. As Erdős and Simonovits pointed out in [46], it implies the general asymptotic description of $\mathrm{ex}(n, \mathcal{L})$.

Theorem 18. *If $\mathcal{L}$ is finite and $\min_{L \in \mathcal{L}} \chi(L) = p > 1$, then*

$$\mathrm{ex}(n, \mathcal{L}) = \left(1 - \frac{1}{p-1}\right)\binom{n}{2} + o(n^2).$$

So this theorem plays a crucial role in extremal graph theory. (For structural generalizations for arbitrary $\mathcal{L}$ see [33,34,115].)

The proof of the Embedding Lemma gives the following quantitative form (see also Frankl-Pach [60], and [118]).

Theorem 19 (Number of copies of H). *Let H be a graph with h vertices and chromatic number p. Let $\beta > 0$ be given and write $\varepsilon = (\beta/6)^h$. If a graph G_n has*

$$e(G_n) > \left(1 - \frac{1}{p-1} + \beta\right)\frac{n^2}{2}$$

then

$$\|H \to G_n\| > \left(\frac{\varepsilon n}{M(\varepsilon)}\right)^h.$$

It is interesting to contrast this with the following peculiar fact observed by Füredi. If a graph has few copies of a sample graph (e.g., few triangles), then they can all be covered by a few edges:

Theorem 20 (Covering copies of H). *For every $\beta > 0$ and sample graph H there is a $\gamma = \gamma(\beta, H) > 0$ such that if G_n is a graph with at most $\gamma n^{v(H)}$ copies of H, then by deleting at most βn^2 edges one can make G_n H-free.*

The above mentioned theorems can be proved directly without the Regularity Lemma, e.g., using sieve-type formulas, see [97,98,48,18].

4 Building Subgraphs

4.1 Building Small Subgraphs

It is well-known that a random graph G_n with fixed edge-density $p > 0$ contains any fixed graph H almost surely (as $n \to \infty$). In some sense this is trivial: we can build up this H vertex by vertex. If we have already fixed ℓ vertices of H then it is easy to find an appropriate $(\ell + 1$-th vertex with the desired connections. The Regularity Lemma (and an application of the Embedding Lemma) achieves the same effect for dense graphs.

4.2 Packing with Small Graphs

The Alon-Yuster Conjecture

The conjecture of Noga Alon and Raphael Yuster [4] generalizes the Hajnal-Szemerédi theorem [73] from covering with cliques to covering with copies of an arbitrary graph H:

Conjecture 21 (Alon-Yuster) *For every graph H there is a constant K such that*

$$\delta(G_n) \geq \left(1 - \frac{1}{\chi(H)}\right) n$$

implies that G_n contains a union of vertex-disjoint copies of H covering all but at most K vertices of G_n.

A simple example in [4] shows that $K = 0$ cannot always be achieved even when $v(H)$ divides $v(G)$. After approximate results of Alon and Yuster [4,5], an exact solution for large n has been given in [95].

Komlós [88] has fine-tuned these covering questions by finding a different degree condition that is (asymptotically) necessary and sufficient. It uses the following quantity:

Definition 5. *For an r-chromatic graph H on h vertices we write $\sigma = \sigma(H)$ for the smallest possible color-class size in any r-coloring of H. The **critical chromatic number** of H is the number*

$$\chi_{cr}(H) = (r - 1)h/(h - \sigma).$$

Theorem 22 (Tiling Turán Theorem [88]). *For every graph H and $\varepsilon > 0$ there is a threshold $n_0 = n_0(H, \varepsilon)$ such that, if $n \geq n_0$ and a graph G_n satisfies the degree condition*

$$\delta(G_n) \geq \left(1 - \frac{1}{\chi_{cr}(H)}\right) n,$$

then G_n contains an H-matching that covers all but at most εn vertices.

4.3 Embedding Trees

So far all embedding questions we discussed dealt with embedding bounded degree graphs H into dense graphs G_n. General Ramsey theory tells us that this cannot be relaxed substantially without putting strong restrictions on the structure of the graph H. (Even for bipartite H, the largest complete bipartite graph $K_{\ell,\ell}$ that a dense graph G_n can be expected to have is for $\ell = O(\log n)$.) A frequently used structural restriction on H is that it is a tree (or a forest). Under this strong restriction even very large graphs H can be embedded into dense graphs G_n.

The two extremal cases are when H is a large star, and when H is a long path. Both cases are precisely and easily handled by classical extremal graph theory (Turán theory or Ramsey theory). The use of the Regularity Lemma makes it possible, in a sense, to reduce the case of general trees H to these two special cases by splitting the tree into "long" and "wide" pieces. After an application of the Regularity Lemma one applies, as always, some classical graph theorem, which in most cases is the König-Hall matching theorem, or the more sophisticated Tutte's theorem (more precisely, the Gallai-Edmonds decomposition).

The Erdős-Sós Conjecture for Trees

Conjecture 23 (Erdős-Sós 1963 [50])   *Every graph on n vertices and more than $(k-1)n/2$ edges contains, as subgraphs, all trees with k edges.*

In other words, if the number of edges in a graph G forces the existence of a k-star, then it also guarantees the existence of every other subtree with k edges. The theorem is known for k-paths (Erdős-Gallai 1959 [40]).

This famous conjecture spurred much activity in graph theory in the last 30 years.

Remark. The assertion is trivial if we are willing to put up with loosing a factor of 2: If G has average degree at least $2k - 2 > 0$, then it has a subgraph G' with $\delta(G') \geq k$, and hence the greedy algorithm guarantees that G' contains all k-trees.

Using an *ad hoc* sparse version of the Regularity Lemma, Ajtai, Komlós, Simonovits and Szemerédi solved the Erdős-Sós conjecture for large n. ([1], in preparation.)

The Loebl Conjecture

In their paper about graph discrepancies P. Erdős, Z. Füredi, M. Loebl and V. T. Sós [39] reduced some questions to the following conjecture of Martin Loebl:

Conjecture 24 (Loebl Conjecture) *If G is a graph on n vertices, and at least $n/2$ vertices have degrees at least $n/2$, then G contains, as subgraphs, all trees with at most $n/2$ edges.*

J. Komlós and V. T. Sós generalized Loebl's conjecture for trees of any size. It says that any graph G contains all trees with size not exceeding the medium degree of G.

Conjecture 25 (Loebl-Komlós-Sós Conjecture) *If G is a graph on n vertices, and at least $n/2$ vertices have degrees greater than or equal to k, then G contains, as subgraphs, all trees with k edges.*

In other words, the condition in the Erdős-Sós conjecture that the *average* degree be greater than $k - 1$, would be replaced here with a similar condition on the *median* degree.

This general conjecture is not easier than the Erdős-Sós conjecture. Large instances of both problems can be attacked with similar methods.

4.4   Embedding Large Bipartite Subgraphs

The following theorem is implicit in Chvátal-Rödl-Szemerédi-Trotter 1983 [24] (according to [2]).

Theorem 26. *For any $\Delta, \beta > 0$ there is a $c > 0$ such that if $e(G_n) > \beta n^2$, then G_n contains as subgraphs all bipartite graphs H with $|V(H)| \leq cn$ and $\Delta(H) \leq \Delta$.*

4.5   Embedding Bounded Degree Spanning Subgraphs

This is probably the most interesting class of embedding problems. Here the proofs (when they exist) are too complicated to quote here, but they follow a general pattern. When embedding H to G (they have the same order now!), we first prepare H by chopping it into (a constant number of) small pieces, then prepare the host graph G by finding a regular partition of G, throw away the usual atypical edges, and define the reduced graph R. Then typically we apply to R the matching theorem (for bipartite H) or the Hajnal-Szemerédi theorem (for r-partite H). At this point, we make an assignment between the small pieces of H and the "regular r-cliques" of the partitioned R. There are two completely different problems left. Make the connections between the r-cliques, and embed a piece of H into an r-clique. The first one is sometimes easy, sometimes very hard, but there is no general recipe to apply here. The second part, however, can typically be handled by referring to the so-called Blow-up Lemma - a new general purpose embedding tool discussed below.

The Pósa-Seymour Conjecture

Paul Seymour conjectured in 1973 that any graph G of order n and minimum degree at least $\frac{k}{k+1}n$ contains the k-th power of a Hamiltonian cycle. For $k = 1$, this is just Dirac's theorem. For $k = 2$, the conjecture was made by Pósa in 1962. Note that the validity of the general conjecture would imply the notoriously hard Hajnal-Szemerédi theorem.

For partial results, see the papers [58,54,55,57,56]. (Fan and Kierstead also announced a proof of the Pósa conjecture if the Hamilton cycle is replaced by Hamilton path.) We do not detail the statements in these papers, since they do not employ the Regularity Lemma.

The Seymour conjecture was proved in [94] for every fixed k and large n.

4.6   The Blow-Up Lemma

Several recent results exist about embedding spanning graphs into dense graphs. Some of the proofs use the following new powerful tool. It basically says that regular pairs behave as complete bipartite graphs from the point of view of embedding bounded degree subgraphs. Note that for embedding spanning subgraphs, one needs all degrees of the host graph to be large. That's why using regular pairs is not sufficient any more, we need super-regular pairs. The Blow-up Lemma plays the same role in embedding spanning graphs H into G as the Embedding Lemma played in embedding smaller graphs H (up to $v(H) < (1 - \varepsilon)v(G)$).

Theorem 27 (Blow-up Lemma - Komlós-Sárközy-Szemerédi 1994 [91]).
Given a graph R of order r and positive parameters δ, Δ, there exists an $\varepsilon > 0$ such that the following holds. Let $n_1, n_2, \ldots, n_r$ be arbitrary positive integers and

let us replace the vertices of R with pairwise disjoint sets $V_1, V_2, \ldots, V_r$ of sizes $n_1, n_2, \ldots, n_r$ (blowing up). We construct two graphs on the same vertex-set $V = \cup V_i$. The first graph $\mathbf{R}$ is obtained by replacing each edge $\{v_i, v_j\}$ of R with the complete bipartite graph between the corresponding vertex-sets V_i and V_j. A sparser graph G is constructed by replacing each edge $\{v_i, v_j\}$ with an (ε, δ)-super-regular pair between V_i and V_j. If a graph H with $\Delta(H) \leq \Delta$ is embeddable into $\mathbf{R}$ then it is already embeddable into G.

The proof of the Blow-up Lemma starts with a probabilistic greedy algorithm, and then uses a König-Hall argument to finish the embedding. The proof of correctness is quite involved, and we will not present it here.

5   Applications in Ramsey Theory

5.1   The Milestone

The following theorem is central in Ramsey theory. It says that the Ramsey number of a bounded degree graph is linear in the order of the graph. In other words, there is a function f such that the graph-Ramsey number $r(H)$ of *any* graph H satisfies $r(H) \leq f(\Delta(H))v(H)$. This was probably the first deep application of the Regularity Lemma, and certainly a milestone in its becoming a standard tool.

Theorem 28 (Chvátal-Rödl-Szemerédi-Trotter 1983 [24]). *For any $\Delta > 0$ there is a $c > 0$ such that if G_n is any n-graph, and H is any graph with $|V(H)| \leq cn$ and $\Delta(H) \leq \Delta$, then either $H \subset G_n$ or $H \subset \overline{G_n}$.*

5.2   Graph-Ramsey

The following more recent theorems also apply the Regularity Lemma.

Theorem 29 (Haxell-Luczak-Tingley 1999 [79]). *Let T_n be a sequence of trees with color-class sizes $a_n \geq b_n$, and let $M_n = \max\{2a_n, a_n + 2b_n\} - 1$ (the trivial lower bound for the Ramsey number $r(T_n)$). If $\Delta(T_n) = o(a_n)$ then $r(T_n) = (1 + o(1))M_n$.*

Theorem 30 (Luczak 1999 [99]). *$R(C_n, C_n, C_n) \leq (3 + o(1))n$ for all even n, and $R(C_n, C_n, C_n) \leq (4 + o(1))n$ for all odd n.*

5.3   Random Ramsey

Given graphs $H_1, \ldots, H_r$ and G, we write $G \to (H_1, \ldots, H_r)$ if for every r-coloring of the edges of G there is an i such that G has a subgraph of color i isomorphic to H_i ('arrow notation'). The typical Ramsey question for random graphs is then the following. What is the threshold edge probability $p = p(n)$ for which $G(n, p) \to (H_1, \ldots, H_r)$ has a probability close to 1.

Rödl and Ruciński [110] answered this in the symmetric case $G_1 = \cdots = G_r$. A first step toward a general solution was taken by Kohayakawa and Kreuter [83] who used the Regularity Lemma to find the threshold when each G_i is a cycle.

6   New Versions of the Regularity Lemma

6.1   The Frieze-Kannan Version

Alan Frieze and Ravi Kannan [62] use a matrix decomposition that can replace the Regularity Lemma in many instances, and creates a much smaller number of parts. The authors describe their approximation algorithm as follows:

Given an $m \times n$ matrix A with entries between -1 and 1, say, and an error parameter ε between 0 and 1, a matrix D is found (by a probabilistic algorithm) which is the sum of $O(1/\varepsilon^2)$ simple rank 1 matrices so that the sum of entries of any submatrix (among the 2^{m+n}) of $(A - D)$ is at most εmn in absolute value. The algorithm takes time dependent only on ε and the allowed probability of failure (but not on m, n).

The rank one matrices in the Frieze-Kannan decomposition correspond to regular pairs in the Regularity Lemma, but the global error term $o(mn)$ is much larger than the one in Szemerédi's theorem. That explains the reasonable sizes ($O(1/\varepsilon^2)$ instead of tower functions).

The decomposition is applied to various standard graph algorithms such as the Max-Cut problem, the Minimum Linear Arrangement problem, and the Maximum Acyclic Subgraph problem, as well as to get quick approximate solutions to systems of linear equations and systems of linear inequalities (Linear Programming feasibility).

The results are also extended from 2-dimensional matrices to r-dimensional matrices.

6.2   A Sparse-Graph Version of the Regularity Lemma

It would be very important to find extensions of the Regularity Lemma for sparse graphs, e.g., for graphs where we assume only that $e(G_n) > cn^{2-\alpha}$ for some positive constants c and α. Y. Kohayakawa [81] and V. Rödl [108] independently proved a version of the Regularity Lemma in 1993 that can be regarded as a Regularity Lemma for sparse graphs. As Kohayakawa puts it: "Our result deals with subgraphs of pseudo-random graphs." He (with co-authors) has also found some interesting applications of this theorem in Ramsey theory and in Anti-Ramsey theory, (see e.g. [75,76,77,78,84,86,83]).

To formulate the Kohayakawa-Rödl Regularity Lemma we need the following definitions.

Definition 6. *A graph $G = G_n$ is (P_0, η)-uniform for a partition P_0 of $V(G_n)$ if for some $p \in [0,1]$ we have*

$$|e_G(U,V) - p|U||V|| \le \eta p|U||V|,$$

whenever $|U|, |V| > \eta n$ and either P_0 is trivial, U, V are disjoint, or U, V belong to different parts of P_0.

Definition 7. *A partition $Q = (C_0, C_1, \ldots, C_k)$ of $V(G_n)$ is (ε, k)-equitable if $|C_0| < \varepsilon n$ and $|C_1| = \cdots = |C_k|$.*

Notation.

$$d_{H,G}(U,V) = \begin{cases} e_H(U,V)/e_G(U,V) & \text{if } e_G(U,V) > 0 \\ 0 & \text{otherwise.} \end{cases}$$

Definition 8. *We call a pair (U,V) (ε, H, G)-regular if for all $U' \subset U$ and $W' \subset W$ with $|U'| \ge \varepsilon|U|$ and $|W'| \ge \varepsilon|W|$, we have*

$$|d_{H,G}(U,W) - d_{H,G}(U',W')| \le \varepsilon.$$

Theorem 31 (Kohayakawa 1993 [81]). *Let ε and $k_0, \ell > 1$ be fixed. Then there are constants $\eta > 0$ and $K_0 > k_0$ with the following properties. For any (P_0, η)-uniform graph $G = G_n$, where $P_0 = (V_i)_i^\ell$ is a partition of $V = V(G)$, if $H \subset G$ is a spanning subgraph of G, then there exists an (ε, H, G)-regular, (ε, k)-equitable partition of V refining P_0, with $k \le k_0 \le K_0$.*

For more information, see Kohayakawa 1997 [82].

7   Algorithmic Questions

The Regularity Lemma is used in two different ways in computer science. Firstly, it is used to prove the existence of some special subconfigurations in given graphs of positive edge-density. Thus by turning the lemma from an existence-theorem into an algorithm one can transform many of the earlier existence results into relatively efficient algorithms. The first step in this direction was made by Alon, Duke, Leffman, Rödl and Yuster [2] (see below). Frieze and Kannan [63] offered an alternative way for constructing a regular partition based on a simple lemma relating non-regularity and largeness of singular values.

In the second type of use, one takes advantage of the fact that the regularity lemma provides a random-like substructure of any dense graph. We know that many algorithms fail on randomlike objects. Thus one can use the Regularity Lemma to prove lower bounds in complexity theory, see e.g., W. Maass and Gy. Turán [72]. One of these randomlike objects is the expander graph, an important structure in Theoretical Computer Science.

7.1   Two Applications in Computer Science

A. Hajnal, W. Maass and Gy. Turán applied the Regularity Lemma to estimate the communicational complexity of certain graph properties [72]. We quote their abstract:

"We prove $\Theta(n \log n)$ bounds for the deterministic 2-way communication complexity of the graph properties CONNECTIVITY, s,t-CONNECTIVITY and BIPARTITENESS. ... The bounds imply improved lower bounds for the VLSI complexity of these decision problems and sharp bounds for a generalized decision tree model that is related to the notion of evasiveness."

Another place where the Regularity Lemma is used in estimating communicational complexity is an (electronic) paper of Pudlák and Sgall [102]. In fact, they only use the (6,3)-problem, i.e., the Ruzsa-Szemerédi theorem.

7.2   An Algorithmic Version of the Regularity Lemma

The Regularity Lemma being so widely applicable, it is natural to ask if for a given graph G_n and given $\varepsilon > 0$ and m one can find an ε-regular partition of G in time polynomial in n. The answer due to Alon, Duke, Lefmann Rödl and Yuster [2] is surprising, at least at first: Given a graph G, we can find regular partitions in polynomially many steps, however, if we describe this partition to someone else, he cannot verify in polynomial time that our partition is really ε-regular: he has better produce his own regular partition. This is formulated below:

Theorem 32. *The following decision problem is co-NP complete: Given a graph G_n with a partition $V_0, V_1, \ldots, V_k$ and an $\varepsilon > 0$. Decide if this partition is ε-regular in the sense guaranteed by the Regularity Lemma.*

Let $Mat(n)$ denote the time needed for the multiplication of two $(0,1)$ matrices of size n.

Theorem 33 (Constructive Regularity Lemma). *For every $\varepsilon > 0$ and every positive integer $t > 0$ there exists an integer $Q = Q(\varepsilon, t)$ such that every graph with $n > Q$ vertices has an ε-regular partition into $k + 1$ classes for some $k < Q$ and such a partition can be found in $O(Mat(n))$ sequential time. The algorithm can be made parallel on an EREW with polynomially many parallel processors, and it will have $O(\log n)$ parallel running time.*

7.3   Counting Subgraphs

Duke, Lefmann and Rödl [30] used a variant of the Regularity Lemma to design an efficient approximation algorithm which, given a labelled graph G on n vertices and a list of all the labelled graphs on k vertices, provides for each graph H in the list an approximation to the number of induced copies of H in G with small total error.

8   Regularity and Randomness

8.1   Extremal Subgraphs of Random Graphs

Answering a question of P. Erdős, L. Babai, M. Simonovits and J. Spencer [6] described the Turán type extremal graphs for random graphs:

> Given an excluded graph L and a probability p, take a random graph R_n of edge-probability p (where the edges are chosen independently) and consider all its subgraphs F_n not containing L. Find the maximum of $e(F_n)$.

Below we formulate four theorems. The first one deals with the simplest case.
    We will use the expression "almost surely" in the sense "with probability $1 - o(1)$ as $n \to \infty$". In this part a p-random graph means a random graph of edge-probability p where the edges are chosen independently.

Theorem 34. *Let $p = 1/2$. If R_n is a p-random graph and F_n is a K_3-free subgraph of R_n containing the maximum number of edges, and B_n is a bipartite subgraph of R_n having maximum number of edges, then $e(B_n) = e(F_n)$. Moreover, F_n is almost surely bipartite.*

Definition 9 (Critical edges). *Given a k-chromatic graph L, an edge e is critical if $L - e$ is $k - 1$-chromatic.*

Many theorems valid for complete graphs were generalized to arbitrary L having critical edges (see e.g., [116]). Theorem 34 also generalizes to every 3-chromatic L containing a critical edge e, and for every probability $p > 0$.

Theorem 35. *Let L be a fixed 3-chromatic graph with a critical edge e (i.e., $\chi(L - e) = 2$). There exists a function $f(p)$ such that if $p \in (0,1)$ is given and $R_n \in \mathbf{G}(p)$, and if B_n is a bipartite subgraph of R_n of maximum size and F_n is an L-free subgraph of maximum size, then*

$$e(B_n) \leq e(F_n) \leq e(B_n) + f(p)$$

almost surely, and almost surely we can delete $f(p)$ edges of F_n so that the resulting graph is already bipartite. Furthermore, there exists a $p_0 < 1/2$ such that if $p \geq p_0$, then F_n is bipartite: $e(F_n) = e(B_n)$.

Theorem 35 immediately implies Theorem 34. The main point in Theorem 35 is that the observed phenomenon is valid not just for $p = 1/2$, but for slightly smaller values of p as well.
    If $\chi(L) = 3$ but we do not assume that L has a critical edge, then we get similar results, having slightly more complicated forms. Here we formulate only some weaker results.

Theorem 36. *Let L be a given 3-chromatic graph. Let $p \in (0,1)$ be fixed and let R_n be a p-random graph. Let $\omega(n) \to 0$ as $n \to \infty$. If B_n is a bipartite subgraph of R_n of maximum size and F_n contains only $\omega(n) \cdot n^{v(L)}$ copies of L and has maximum size under this condition, then almost surely*

$$e(B_n) \le e(F_n) \le e(B_n) + o(n^2)$$

and we can delete $o(n^2)$ edges of F_n so that the resulting graph is already bipartite.

The above results also generalize to r-chromatic graphs L.

Some strongly related important results are hidden in the paper of Haxell, Kohayakawa and Łuczak [77].

8.2   Quasirandomness

Quasi-random structures have been investigated by several authors, among others, by Thomason [129], Chung, Graham, Wilson, [23]. For graphs, Simonovits and T. Sós [118] have shown that quasi-randomness can also be characterized by using the Regularity Lemma. Fan Chung [22] generalized their results to hypergraphs.

Let $N_G^*(L)$ and $N_G(L)$ denote the number of induced and not necessarily induced copies of L in G, respectively. Let $S(x,y)$ be the set of vertices joined to both x and y in the same way. First we formulate a theorem of Chung, Graham, and Wilson, in a shortened form.

Theorem 37 (Chung-Graham-Wilson [23]). *For any graph sequence (G_n) the following properties are equivalent:*
$\mathbf{P}_1(\nu)$:   *for fixed ν, for all graphs H_ν*

$$N_G^*(H_\nu) = (1 + o(1))n^\nu 2^{-\binom{\nu}{2}}.$$

$\mathbf{P}_2(t)$:   *Let C_t denote the cycle of length t. Let $t \ge 4$ be even.*

$$e(G_n) \ge \frac{1}{4}n^2 + o(n^2) \qquad \text{and} \qquad N_G(C_t) \le \left(\frac{n}{2}\right)^t + o(n^t).$$

$\mathbf{P}_5$:   *For each subset $X \subset V$, $|X| = \lfloor \frac{n}{2} \rfloor$ we have $e(X) = \left(\frac{1}{16}n^2 + o(n^2)\right)$.*
$\mathbf{P}_6$: $\sum_{x,y \in V} \left| |S(x,y)| - \frac{n}{2} \right| = o(n^3)$.

Graphs satisfying these properties are called **quasirandom**. Simonovits and T. Sós formulated a graph property which proved to be equivalent with the above properties.
$\mathbf{P}_S$:   For every $\varepsilon > 0$ and κ there exist two integers, $k(\varepsilon, \kappa)$ and $n_0(\varepsilon, \kappa)$ such that for $n \ge n_0$, G_n has a regular partition with parameters ε and κ and k classes $U_1, \dots, U_k$, with $\kappa \le k \le k(\varepsilon, \kappa)$, so that

$$(U_i, U_j) \text{ is } \varepsilon - \text{regular, and} \qquad \left| d(U_i, U_j) - \frac{1}{2} \right| < \varepsilon$$

holds for all but at most εk^2 pairs (i, j), $1 \le i, j \le k$.

It is easy to see that if (G_n) is a random graph sequence of probability $1/2$, then $\mathbf{P}_S$ holds for (G_n), almost surely. Simonovits and T. Sós [118] proved that $\mathbf{P}_S$ is a quasi-random property, i.e. $\mathbf{P}_S \Longleftrightarrow \mathbf{P}_i$ for all the above properties P_i.

8.3   Hereditarily Extended Properties

Randomness is a hereditary property: large subgraphs of random graphs are fairly randomlike. In [119] and [120] Simonovits and T. Sós proved that some properties which are not quasi random, become quasirandom if one extends them to hereditary properties. This "extension" means that the properties are assumed not only for the whole graph but for all sufficiently large subgraphs. Their most interesting results were connected with counting some small subgraphs $L \subseteq G_n$.

Obviously, $\mathbf{P}_1(\nu)$ of Theorem 37 says that the graph G_n contains each subgraph with the same frequency as a random graph.

Let $\nu = v(L)$, $E = e(L)$. Denote by $\beta_L(p)$ and $\gamma_L(p)$ the "densities" of **labelled induced** and **labelled not necessarily induced** copies of L in a p-random graph:

$$\beta_L(p) = p^E(1-p)^{\binom{\nu}{2}-E} \quad \text{and} \quad \gamma_L(p) = p^E. \tag{1}$$

Theorem 38 (Simonovits-Sós). *Let L_ν be a fixed sample-graph, $e(L) > 0$, $p \in (0,1)$ be fixed. Let (G_n) be a sequence of graphs. If (for every sufficiently large n) for every induced $F_h \subseteq G_n$,*

$$\mathbf{N}(L_\nu \subseteq F_h) = \gamma_L(p)h^\nu + o(n^\nu), \tag{2}$$

then (G_n) is p-quasi-random.

Observe that in (2) we used $o(n^\nu)$ instead of $o(h^\nu)$, i.e., for small values of h we allow a relatively much larger error-term. As soon as $h = o(n)$, this condition is automatically fulfilled.

For "Induced Copies" the situation is much more involved, because of the lack of monotonicity. Below we shall always exclude $e(L_\nu) = 0$ and $e(\overline{L}_\nu) = 0$. One would like to know if for given (L_ν, p) the following is true or not:

(#)   Given a sample graph L_ν and a probability p, if for a graph graph sequence (G_n) for every induced subgraph F_h of G_n

$$\mathbf{N}^*(L_\nu \subseteq F_h) = \beta_L(p)h^\nu + o(n^\nu) \tag{3}$$

then (G_n) is p-quasi-random.

(#) is *mostly false* in this form, for two reasons:
- the probabilities are in **conjugate pairs**;
- There may occur strange **algebraic coincidences**.

Clearly, $\beta_L(p)$ (in (1)) is a function of p which is monotone increasing in $\left[0, e(L_\nu)/\binom{\nu}{2}\right]$, monotone decreasing in $\left[e(L_\nu)/\binom{\nu}{2}, 1\right]$ and vanishes in $p = 0$

and in $p = 1$. For every $p \in \left(0, e(L_\nu)/\binom{\nu}{2}\right)$ there is a *unique* probability $\overline{p} \in \left(e(L_\nu)/\binom{\nu}{2}, 1\right)$ yielding the same expected value. Therefore the hereditarily assumed number of induced copies does not determine the probability uniquely, unless $p = e(L_\nu)/\binom{\nu}{2}$. Given a graph L_ν, the probabilities p and $\overline{p}$ are called **conjugate** if $\beta_L(p) = \beta_L(\overline{p})$. We can mix two such sequences: (G_n) obviously satisfies (3) if

$(*)$ (G_n) is the union of a p-quasi random graph sequence and
             a $\overline{p}$-quasi random graph sequence.

One can create such sequences for P_3 or its complementary graph. Simonovits and Sós think that there are no other real counterexamples:

Conjecture 39 *Let L_ν be fixed, $\nu \geq 4$ and $p \in (0,1)$. Let (G_n) be a graph sequence satisfying (3). Then (G_n) is the union of two sequences, one being p-quasi-random, the other $\overline{p}$-quasi-random (where one of these two sequences may be finite, or even empty).*

To formulate the next Simonovits-Sós theorem we use

Construction 40 (Two class generalized random graph) *Define the graph $G_n = G(V_1, V_2, p, q, s)$ as follows: $V(G_n) = V_1 \cup V_2$. We join independently the pairs in V_1 with probability p, in V_2 with probability q and the pairs (x, y) for $x \in V_1$ and $y \in V_2$ with probability s.*

Theorem 41 (Two-class counterexample). *If there is a sequence (G_n) which is a counterexample to Conjecture 39 for a fixed sample graph L and a probability $p \in (0,1)$, then there is also a 2-class generalized random counterexample graph sequence of form $G_n = G(V_1, V_2, p, q, s)$ with $|V_1| \approx n/2$, $p \in (0,1)$, $s \neq p$. (Further, either $q = p$ or $q = \overline{p}$.)*

This means that if there are counterexamples then those can be found by solving some systems of algebraic equations. The proof of this theorem heavily uses the regularity lemma.

Theorem 42. *If L_ν is **regular**, then Conjecture 39 holds for L_ν and any $p \in (0,1)$.*

For some further results of Simonovits and T. Sós for induced subgraphs see [120].

References

[1] M. Ajtai, J. Komlós, M. Simonovits and E. Szemerédi, Solution of the Erdős-Sós Conjecture, in preparation.
[2] N. Alon, R. Duke, H. Leffman, V. Rödl, R. Yuster, The algorithmic aspects of the regularity lemma, FOCS 33 (1992), 479-481, Journal of Algorithms 16 (1994), 80-109.

[3] N. Alon, E. Fischer, 2-factors in dense graphs, Discrete Math.

[4] N. Alon, R. Yuster, Almost H-factors in dense graphs, Graphs and Combinatorics 8 (1992), 95-102.

[5] N. Alon, R. Yuster, H-factors in dense graphs, J. Combinatorial Theory **B66** (1996), 269-282.

[6] L. Babai, M. Simonovits, J. Spencer, Extremal subgraphs of random graphs, Journal of Graph Theory **14** (1990), 599-622.

[7] V. Bergelson, A. Leibman, Polynomial extension of van der Waerden's and Szemerédi's theorem.

[8] B. Bollobás, Extremal graph theory, Academic Press, London (1978).

[9] Béla Bollobás, The work of William Timothy Gowers, Proceedings of the International Congress of Mathematicians, Vol. I (Berlin, 1998), Doc. Math. 1998, Extra Vol. I, 109-118 (electronic).

[10] B. Bollobás, P. Erdős, On a Ramsey-Turán type problem, Journal of Combinatorial Theory B21 (1976), 166-168.

[11] B. Bollobás, P. Erdős, M. Simonovits, E. Szemerédi, Extremal graphs without large forbidden subgraphs, Annals of Discrete Mathematics 3 (1978), 29-41, North-Holland.

[12] B. Bollobás, A. Thomason, The structure of hereditary properties and colourings of random graphs. Combinatorica **20** (2000), 173-202.

[13] W. G. Brown, P. Erdős, M. Simonovits, Extremal problems for directed graphs, Journal of Combinatorial Theory **B15** (1973), 77-93.

[14] W. G. Brown, P. Erdős, M. Simonovits, Inverse extremal digraph problems, Colloq. Math. Soc. J. Bolyai **37** (Finite and Infinite Sets), Eger (Hungary) 1981, Akad. Kiadó, Budapest (1985), 119-156.

[15] W. G. Brown, P. Erdős, M. Simonovits, Algorithmic solution of extremal digraph problems, Transactions of the American Math. Soc. **292/2** (1985), 421-449.

[16] W. G. Brown, P. Erdős, V. T. Sós, Some extremal problems on r-graphs, New directions in the theory of graphs (Proc. Third Ann Arbor Conf., Univ. Michigan, Ann Arbor, Mich, 1971), 53-63. Academic Press, New York, 1973.

[17] W. G. Brown, P. Erdős, V. T. Sós, On the existence of triangulated spheres in 3-graphs, and related problems, Period. Math. Hungar. **3** (1973), 221-228.

[18] W. G. Brown, M. Simonovits, Digraph extremal problems, hypergraph extremal problems, and densities of graph structures, Discrete Mathematics 48 (1984), 147-162.

[19] S. Burr, P. Erdős, P. Frankl, R. L. Graham, V. T. Sós, Further results on maximal antiramsey graphs, Proc. Kalamazoo Combin. Conf. (1989), 193-206.

[20] S. Burr, P. Erdős, R. L. Graham, V. T. Sós, Maximal antiramsey graphs and the strong chromatic number (The nonbipartite case) Journal of Graph Theory **13** (1989), 163-182.

[21] L. Caccetta, R. Häggkvist, On diameter critical graphs, Discrete Mathematics **28** (1979), 223-229.

[22] Fan R. K. Chung, Regularity lemmas for hypergraphs and quasi-randomness, Random Structures and Algorithms **2** (1991), 241-252.

[23] F. R. K. Chung, R. L. Graham, R. M. Wilson, Quasi-random graphs, Combinatorica **9** (1989), 345-362.

[24] V. Chvátal, V. Rödl, E. Szemerédi, W. T. Trotter Jr., The Ramsey number of a graph with bounded maximum degree, Journal of Combinatorial Theory B34 (1983), 239-243.

[25] V. Chvátal, E. Szemerédi, On the Erdős-Stone theorem, Journal of the London Math. Soc. **23** (1981), 207-214.

[26] V. Chvátal, E. Szemerédi, Notes on the Erdős-Stone theorem, Combinatorial Mathematics, Annals of Discrete Mathematics **17** (1983), (Marseille-Luminy, 1981), 183-190, North-Holland, Amsterdam-New York, 1983.

[27] K. Corrádi, A. Hajnal, On the maximal number of independent circuits in a graph, Acta Math. Acad. Sci. Hung. **14** (1963), 423-439.

[28] W. Deuber, Generalizations of Ramsey's theorem, Proc. Colloq. Math. Soc. János Bolyai **10** (1974), 323-332.

[29] G. A. Dirac, Some theorems on abstract graphs, Proc. London Math. Soc. **2** (1952), 68-81.

[30] Duke, Richard A., Hanno Lefmann, Hanno, Vojtěch Rödl, A fast approximation algorithm for computing the frequencies of subgraphs in a given graph, SIAM J. Comput. 24 (1995), 598-620.

[31] R. A. Duke, V. Rödl, On graphs with small subgraphs of large chromatic number, Graphs Combin. **1** (1985), 91-96.

[32] R. A. Duke, V. Rödl, The Erdős-Ko-Rado theorem for small families, J. Combin. Theory Ser. **A65** (1994), 246-251.

[33] P. Erdős, Some recent results on extremal problems in graph theory, Results, International Symposium, Rome (1966), 118-123.

[34] P. Erdős, On some new inequalities concerning extremal properties of graphs, Theory of Graphs, Proc. Coll. Tihany, Hungary (P. Erdős and G. Katona eds.) Acad. Press N. Y. (1968), 77-81.

[35] P. Erdős, On some extremal problems on r-graphs, Discrete Mathematics **1** (1971), 1-6.

[36] P. Erdős, Some old and new problems in various branches of combinatorics, Proc. 10th Southeastern Conf. on Combinatorics, Graph Theory and Computation, Boca Raton (1979) Vol I., Congressus Numerantium **23** (1979), 19-37.

[37] P. Erdős, On the combinatorial problems which I would most like to see solved, Combinatorica **1** (1981), 25-42.

[38] P. Erdős, P. Frankl, V. Rödl, The asymptotic number of graphs not containing a fixed subgraph and a problem for hypergraphs having no exponent, Graphs and Combinatorics **2** (1986), 113-121.

[39] P. Erdős, Z. Füredi, M. Loebl, V. T. Sós, Studia Sci. Math. Hung. **30** (1995), 47-57. (Identical with the book Combinatorics and its applications to regularity and irregularity of structures, W. A. Deuber and V. T. Sós eds., Akadémiai Kiadó, 47-58.)

[40] P. Erdős, T. Gallai, On maximal paths and circuits of graphs, Acta Math. Acad. Sci. Hung. **10** (1959), 337-356.

[41] P. Erdős, A. Hajnal, On complete topological subgraphs of certain graphs, Annales Univ. Sci. Budapest **7** (1969), 193-199.

[42] P. Erdős, A. Hajnal, L. Pósa, Strong embedding of graphs into colored graphs, Proc. Colloq. Math. Soc. János Bolyai **10** (1975), 585-595.

[43] P. Erdős, A. Hajnal, M. Simonovits, V. T. Sós, E. Szemerédi, Turán-Ramsey theorems and simple asymptotically extremal structures, Combinatorica **13** (1993), 31-56.

[44] P. Erdős, A. Hajnal, M. Simonovits, V. T. Sós, E. Szemerédi, Turán-Ramsey theorems for K_p-stability numbers, Proc. Cambridge, also in Combinatorics, Probability and Computing **3** (1994) (P. Erdős birthday meeting), 297-325.

[45] P. Erdős, A. Hajnal, V. T. Sós, E. Szemerédi, More results on Ramsey-Turán type problems, Combinatorica **3** (1983), 69-81.

[46] P. Erdős, M. Simonovits, A limit theorem in graph theory, Studia Sci. Math. Hung. **1** (1966), 51-57.

[47] P. Erdős, M. Simonovits, The chromatic properties of geometric graphs, Ars Combinatoria **9** (1980), 229-246.

[48] P. Erdős, M. Simonovits, Supersaturated graphs and hypergraphs, Combinatorica **3** (1983), 181-192.

[49] P. Erdős, M. Simonovits, How many colours are needed to colour every pentagon of a graph in five colours? (to be published).

[50] P. Erdős, V. T. Sós, The tree conjecture, Mentioned in P. Erdős, Extremal problems in graph theory, Theory of graphs and its applications, Proc. of the Symposium held in Smolenice in June 1963, 29-38.

[51] P. Erdős, V. T. Sós, Some remarks on Ramsey's and Turán's theorem, Combinatorial theory and its applications, II (Proc. Colloq., Balatonfüred, 1969), 395-404, North-Holland, Amsterdam, 1970.

[52] P. Erdős, A. H. Stone, On the structure of linear graphs, Bull. Amer. Math. Soc. **52** (1946), 1089-1091.

[53] P. Erdős, P. Turán, On some sequences of integers, J. London Math. Soc. **11** (1936), 261-264.

[54] G. Fan, R. Häggkvist, The square of a hamiltonian cycle, SIAM J. Disc. Math.

[55] G. Fan, H. A. Kierstead, The square of paths and cycles, Journal of Combinatorial Theory B63 (1995), 55-64.

[56] G. Fan, H. A. Kierstead, The square of paths and cycles II.

[57] R. J. Faudree, R. J. Gould, M. Jacobson, On a problem of Pósa and Seymour.

[58] R. J. Faudree, R. J. Gould, M. S. Jacobson, R. H. Schelp, Seymour's conjecture, Advances in Graph Theory (V. R. Kulli ed.), Vishwa International Publications (1991), 163-171.

[59] P. Frankl, Z. Füredi, Exact solution of some Turán-type problems, Journal of Combinatorial Theory A45 (1987), 226-262.

[60] P. Frankl, J. Pach, An extremal problem on K_r-free graphs, Journal of Graph Theory **12** (1988), 519-523.

[61] P. Frankl, V. Rödl, The Uniformity Lemma for hypergraphs, Graphs and Combinatorics **8** (1992), 309-312.

[62] Alan Frieze, Ravi Kannan, Quick approximation to matrices and applications, Combinatorica **19** (1999), 175-220.

[63] Alan Frieze, Ravi Kannan, A simple algorithm for constructing Szemerédi's regularity partition, Electron. J. Combin. **6** (1999), Research Paper 17 (electronic).

[64] Z. Füredi, Turán type problems, in Surveys in Combinatorics (1991), Proc. of the 13th British Combinatorial Conference, (A. D. Keedwell ed.) Cambridge Univ. Press. London Math. Soc. Lecture Note Series 166 (1991), 253-300.

[65] Z. Füredi, The maximum number of edges in a minimal graph of diameter 2, Journal of Graph Theory **16** (1992), 81-98.

[66] H. Fürstenberg, Ergodic behavior of diagonal measures and a theorem of Szemerédi on arithmetic progressions, Journal d'Analyse Math. 31 (1977), 204-256.

[67] H. Fürstenberg, A polynomial Szemerédi theorem, Combinatorics, Paul Erdős is eighty, Vol. 2 (Keszthely, 1993), 1-16, Bolyai Soc. Math. Stud., 2, János Bolyai Math. Soc., Budapest, 1996

[68] H. Fürstenberg, Y. Katznelson, Idempotents in compact semigroups and Ramsey theory, Israel Journal of Mathematics **68** (1989), 257-270.

[69] H. Fürstenberg, Y. Katznelson, A density version of the Hales-Jewett theorem, Journal d'Analyse Math. **57** (1991), 64-119.

[70] W. T. Gowers, Lower bounds of tower type for Szemerédi's uniformity lemma, Geom. Funct. Anal. **7** (1997), 322-337.

[71] R. L. Graham, B. L. Rothschild, J. Spencer, Ramsey Theory, Wiley Interscience, Series in Discrete Mathematics (1980).

[72] A. Hajnal, W. Maass, Gy. Turán, On the communication complexity of graph properties, 20th STOC, Chicago (1988), 186-191.

[73] A. Hajnal, E. Szemerédi, Proof of a conjecture of Erdős, Combinatorial Theory and its Applications vol. II (P. Erdős, A. Rényi and V. T. Sós eds.), Colloq. Math. Soc. J. Bolyai 4, North-Holland, Amsterdam (1970), 601-623.

[74] P. E. Haxell, Y. Kohayakawa, The size-Ramsey number of trees, Israel J. Math. 89 (1995), 261-274.

[75] P. E. Haxell, Y. Kohayakawa, On an anti-Ramsey property of Ramanujan graphs, Random Structures and Algorithms 6 (1995), 417-431.

[76] P. E. Haxell, Y. Kohayakawa, T. Łuczak, The induced size-Ramsey number of cycles, Combinatorics, Probability and Computing 4 (1995), 217-239.

[77] P. E. Haxell, Y. Kohayakawa, T. Łuczak, Turán's extremal problem in random graphs: forbidding even cycles, Journal of Combinatorial Theory B64 (1995), 273-287.

[78] P. E. Haxell, Y. Kohayakawa, T. Łuczak, Turán's extremal problem in random graphs: forbidding odd cycles, Combinatorica 16 (1996), 107-122.

[79] P. E. Haxell, T. Luczak, P. W. Tingley, Ramsey Numbers for Trees of Small Maximum Degree.

[80] D. R. Heath-Brown, Integer sets containing no arithmetic progressions, J. London Math. Soc. 35 (1987), 385-394.

[81] Y. Kohayakawa, The Regularity Lemma of Szemerédi for sparse graphs, manuscript, August 1993.

[82] Y. Kohayakawa, Szemerédi's regularity lemma for sparse graphs, Foundations of computational mathematics (Rio de Janeiro) (1997), 216-230, Springer, Berlin.

[83] Y. Kohayakawa, B. Kreuter, Threshold functions for asymmetric Ramsey properties involving cycles, Random Structures Algorithms 11 (1997), 245-276.

[84] Y. Kohayakawa, T. Łuczak, V. Rödl, Arithmetic progressions of length three in subsets of a random set, Acta Arithmetica, 75 (1996), 133-163.

[85] Y. Kohayakawa, T. Łuczak, V. Rödl, Arithmetic progressions of length three in subsets of a random set, Acta Arith. 75 (1996), 133-163.

[86] Y. Kohayakawa, T. Łuczak, V. Rödl, On K^4-free subgraphs of random graphs, Combinatorica 17 (1997), 173-213.

[87] J. Komlós, The blow-up lemma, Recent trends in combinatorics (Mátraháza, 1995), Combin. Probab. Comput. 8 (1999), 161-176.

[88] J. Komlós, Tiling Turán Theorems, Combinatorica 20 (2000), 203-218.

[89] J. Komlós, G. N. Sárközy, E. Szemerédi, Proof of a packing conjecture of Bollobás, AMS Conference on Discrete Mathematics, DeKalb, Illinois (1993), Combinatorics, Probability and Computing 4 (1995), 241-255.

[90] J. Komlós, G. N. Sárközy, E. Szemerédi, On the square of a Hamiltonian cycle in dense graphs, Proceedings of the Seventh International Conference on Random Structures and Algorithms (Atlanta, GA, 1995), Random Structures and Algorithms 9 (1996), 193-211.

[91] J. Komlós, G. N. Sárközy, E. Szemerédi, Blow-up Lemma, Combinatorica 17 (1997), 109-123.

[92] J. Komlós, G. N. Sárközy, E. Szemerédi, An algorithmic version of the blow-up lemma, Random Structures Algorithms 12 (1998), 297-312.

[93] J. Komlós, G. N. Sárközy, E. Szemerédi, On the Pósa-Seymour conjecture, J. Graph Theory 29 (1998), 167-176.

[94] János Komlós, Gábor Sárközy, Endre Szemerédi, Proof of the Seymour conjecture for large graphs, Ann. Comb. **2** (1998), 43-60.

[95] J. Komlós, G. N. Sárközy, E. Szemerédi, Proof of the Alon-Yuster conjecture, Random Structures and Algorithms.

[96] J. Komlós, M. Simonovits, Szemerédi's regularity lemma and its applications in graph theory, Bolyai Society Mathematical Studies 2, Combinatorics, Paul Erdős is Eighty (Volume 2) (D. Miklós, V. T. Sós, T. Szőnyi eds.), Keszthely (Hungary) (1993), Budapest (1996), 295-352.

[97] L. Lovász, M. Simonovits, On the number of complete subgraphs of a graph I, Proc. Fifth British Combin. Conf. Aberdeen (1975), 431-442.

[98] L. Lovász, M. Simonovits, On the number of complete subgraphs of a graph II, Studies in Pure Math (dedicated to the memory of P. Turán), Akadémiai Kiadó and Birkhäuser Verlag (1983), 459-495.

[99] T. Luczak, $R(C_n, C_n, C_n) \leq (4 + o(1))n$, J. Combin. Theory **B75** (1999), 174-187.

[100] Luczak, Tomasz; Rödl, Vojtěch; Szemerédi, Endre, Partitioning two-coloured complete graphs into two monochromatic cycles, Combin. Probab. Comput. **7** (1998), 423-436.

[101] J. Nešetřil, V. Rödl, Partition theory and its applications, in Surveys in Combinatorics (Proc. Seventh British Combinatorial Conf., Cambridge, 1979), pp. 96-156, (B. Bollobás ed.), London Math. Soc. Lecture Notes Series, Cambridge Univ. Press, Cambridge-New York, 1979.

[102] P. Pudlák, J. Sgall, An upper bound for a communication game, related to time-space tradeoffs, Electronic Colloquium on Computational Complexity, TR 95-010, (1995).

[103] K. F. Roth, On certain sets of integers (II), J. London Math. Soc. **29** (1954), 20-26.

[104] K. F. Roth, Irregularities of sequences relative to arithmetic progressions (III), Journal of Number Theory **2** (1970), 125-142.

[105] K. F. Roth, Irregularities of sequences relative to arithmetic progressions (IV), Periodica Math. Hung. **2** (1972), 301-326.

[106] V. Rödl, A generalization of Ramsey Theorem and dimension of graphs, Thesis, 1973, Charles Univ. Prague; see also: A generalization of Ramsey Theorem for graphs, hypergraphs and block systems, Zielona Gora (1976), 211-220.

[107] V. Rödl, On universality of graphs with uniformly distributed edges, Discrete Mathematics **59** (1986), 125-134.

[108] V. Rödl, Sparse Regularity, Personal communication.

[109] V. Rödl, A. Ruciński, Random graphs with monochromatic triangles in every edge coloring, Random Structures and Algorithms **5** (1994), 253-270.

[110] V. Rödl, A. Ruciński, Threshold functions for Ramsey properties, J. Amer. Math. Soc. **8** (1995), 917-942.

[111] I. Z. Ruzsa, E. Szemerédi, Triple systems with no six points carrying three triangles, Combinatorics (Keszthely, 1976), **18** (1978), Vol. II., 939-945, North-Holland, Amsterdam-New York.

[112] G. N. Sárközy, Fast parallel algorithm for finding Hamiltonian cycles and trees in graphs.

[113] P. Seymour, Problem section, Combinatorics: Proceedings of the British Combinatorial Conference 1973 (T. P. McDonough and V. C. Mavron eds.), Cambridge University Press (1974), 201-202.

[114] A. F. Sidorenko, Boundedness of optimal matrices in extremal multigraph and digraph problems, Combinatorica **13** (1993), 109-120.

[115] M. Simonovits, A method for solving extremal problems in graph theory, Theory of graphs, Proc. Coll. Tihany (1966) (P. Erdős and G. Katona eds.), Acad. Press, N.Y. (1968), 279-319.

[116] M. Simonovits, Extremal graph problems with symmetrical extremal graphs, additional chromatic conditions, Discrete Mathematics **7** (1974), 349-376.

[117] M. Simonovits, Extremal graph theory, Selected Topics in Graph Theory (L. Beineke and R. Wilson eds.) Academic Press, London, New York, San Francisco (1985), 161-200.

[118] M. Simonovits, V. T. Sós, Szemerédi's partition and quasirandomness, Random Structures and Algorithms **2** (1991), 1-10.

[119] M. Simonovits and V. T. Sós, Hereditarily extended properties, quasi-random graphs and not necessarily induced subgraphs, Combinatorica **17** (1997), 577-596.

[120] M. Simonovits and V. T. Sós, Hereditarily extended properties, quasi-random graphs and induced subgraphs, to be published.

[121] M. Simonovits and V. T. Sós, Ramsey-Turán theory, Proc. Prague Meeting, Fifth Czech-Slovak International Symposia, Discrete Mathematics, to appear.

[122] V. T. Sós, On extremal problems in graph theory, Proc. Calgary International Conf. on Combinatorial Structures and their Application, Gordon and Breach, N. Y. (1969), 407-410.

[123] V. T. Sós, Interaction of Graph Theory and Number Theory, *in* Proc. Conf. Paul Erdős and His Mathematics, Budapest, 1999. Springer Verlag, 2001.

[124] E. Szemerédi, On sets of integers containing no four elements in arithmetic progression, Acta Math. Acad. Sci. Hung. **20** (1969), 89-104.

[125] E. Szemerédi, On graphs containing no complete subgraphs with 4 vertices (in Hungarian), Matematikai Lapok **23** (1972), 111-116.

[126] E. Szemerédi, On sets of integers containing no k elements in arithmetic progression, Acta Arithmetica **27** (1975), 199-245.

[127] E. Szemerédi, Regular partitions of graphs, Colloques Internationaux C.N.R.S. Nº 260 - Problèmes Combinatoires et Théorie des Graphes, Orsay (1976), 399-401.

[128] E. Szemerédi, Integer sets containing no arithmetic progressions, Acta Math. Acad. Sci. Hung. **56** (1990), 155-158.

[129] A. Thomason, Pseudo-random graphs, in Proc. of Random Graphs, Poznán (1985) (M. Karoński ed.), Annals of Discr. Math. (North-Holland) 33 (1987), 307-331. See also: Dense expanders and bipartite graphs, Discrete Mathematics 75 (1989), 381-386.

[130] Pál Turán, On an extremal problem in graph theory (in Hungarian), Matematikai és Fizikai Lapok **48** (1941), 436-452.

[131] B. L. Van der Waerden, Beweis einer Baudetschen Vermutung, Nieuw Archief voor Wiskunde **15** (1927), 212-216.

[132] R. Yuster, The number of edge colorings with no monochromatic triangle, J. Graph Theory **21** (1996), 441-452.

Modeling Data and Objects: An Algebraic View Point

Kazem Lellahi

LIPN, UPRES-A 7030 C.N.R.S
Université Paris 13, Institut Galilée,
93430 Villetaneuse
France
kl@lipn.univ-paris13.fr

Abstract. This paper proposes an algebraic semantics approach for data and object modeling. The approach captures the main concepts of object systems, namely: class, method, object identity, inheritance, overriding, overloading, late and early binding, collection types and persistence objects. The proposed model follows the algebraic aspects of relational database tradition; that is, the clear separation between schema, types (or domains), instance and query. For this reason it is enable to support an algebraic query language in the style of the relational algebra. Our approach also provides a rigorous mathematical treatment of null values in the object-oriented systems.

1 Introduction

A traditionally database instance is a collection of records. The record fields are called *attributes*. The type of a record's field is either basic as in the pure relational model [3], or a set record type as in the nested relational model [29], [28], or any combination of set and record types as in complex object models [2]. Some of these models also support null values, i.e. the value of a record's field may be undefined [32], [33], [26]. However, most of these models do not reflect the semantics of real world objects. On the other hand, object-oriented models claim to overcome this semantics problem.

In object-oriented models [17], [19], [8], [1] each record, called an object in the trade, has a unique identification, and records may have other fields which are calculations or *methods*. The value of a calculation field is obtained by evaluating a code which is often written in a programming language. Objects are organized in *classes* which are named collections of objects of the same kind. Classes are organized in an *inheritance hierarchy*. Another important aspect of object modeling is overloading, which is the possibility of giving the same name to several methods.

In fact most object-oriented database systems are embedded in a programming language which is enriched with database functionality such as persistence and query processing features. In the absence of a standard formal object data model, various models and various query languages have been proposed [4], [18],

G.B. Khosrovshahi et al. (Eds.): Theoretical Aspects of Computer Science, LNCS 2292, pp. 113–147, 2002.
© Springer-Verlag Berlin Heidelberg 2002

[16], [14], [15], [7]. The ODMG group carried out a task of standardization and proposed an object data language ODL, and an object query language OQL [9]. However, no formal model has emerged with the same authority as the relational model, nor any algebraic query language with the same elegance as the relational algebra. We think that the existing proposals contain enough material for defining a formal model and an algebraic query language. In this paper we intend to move towards such a formal model, and to provide rigorous mathematical definitions for all the concepts of object-oriented models, namely : attributes, methods, classes, object identity, inheritance, overriding, overloading, dynamic and static binding, object identity, collection, persistence, query specification and query evaluation.

The originality of our formalism is that it follows the relational database tradition, that is to say, the clear separation between schema, domain, instance and query. To carry through this objective we consider an object-oriented database as a collection of partial functions. Each function has at least one parameter which represents an object. On the conceptual level other parameters are less important. Therefore, inspired by the standard Curryfication technique from functional programming, we assume that each function has arity 1 and its unique parameter represents the unique identification of an object (i.e. object identity). The result of each function on an object is a calculation or a value of some type. We distinguish between two kinds of calculation : calculation with a result and calculation with side-effects. The former do not change the state whereas the latter may do. For example, an *update* procedure is a calculation with side-effects.

With this point of view, a class is the collection of partial functions with the same domain type : the class name. Types are expressions obtained from basic types, class names and constructors. In this paper we develop the approach with two main constructors *set* and $\otimes$. But other similar constructors may be envisaged. Our product has two peculiarities. In one hand it is not a field named product as is the record type. It is, however, endowed with projections fst and snd as in some functional programming language. On the other hand, we give a special semantics for $\otimes$ which is not that of the usual cartesian product. This semantics is more suitable for a rigorous treatment of null and undefined values in the object paradigm. Based on this type system we define formally what is an object-oriented database schema. The definition of object-oriented database schema follows that of relational database schema but includes concepts of inheritance and behavior. We show how classical concepts of objects can be translated into this model. Inheritance serves to organize classes under a hierarchy. Usually, in object systems type are also organized into a hierarchy, but often independent of class hierarchy. It is well known that the interference between inheritance, subtyping and overloading may cause inconsistency design problems, and render ambiguous derivations of methods and attributes. Usually ad-hoc solutions are offered for solving the problem. The solution often consists of restricting inheritance and overloading. For instance, in Java or in C++ inheritance is simple and an overridden method have to preserve argument types of the original method. In our approach there are no restrictions on inheritance, and overriding is less restrictive. In contrast, our subtyping relation is an extension to types of our inheritance relations. We believe that this subtyping is more convenient to database context. As a result we can propose a formal rigorous solution to the derivation problems even with multiple inheritance and with our less restrictive overloading. We define rigorously what means a hierarchically consistent schema and we provide procedures for checking such hierarchical consistency. Then we give a formal definition of a database instance over a hierarchically consistent schema and propose an algebraic query language in the style of the relational model. The instance includes a classification function serving to classify stored objects into classes. We show how inheritance in a database context enables to identify objects in different classes. Within a database instance a value have a static type and a dynamic type. These are the counterpart of analogous concepts in object programming languages.

A query expression can contains attribute names as well as method names and other kind of calculations over objects. However, calculations used in a query have to be calculations without side effects. Evaluation of a query takes into account the hierarchy of objects and values. Our algebraic operations of the query language generalize those of the relational model and its extensions as well as those of standard languages like SQL and OQL.

To integrate methods and calculations in queries, the query language have to be assisted either by one or several programming languages, or the data model have to include programming constructions. Most commercial language, for example SQL, have both facilities (pro*C/C++, PL/SQL, ...). We endowed our model with the kernel of a language. This is a pure functional and variable free language. Its variable free aspect permits to see it as an algebraic language. In fact it is an algebra for partial functions. The algebra is freely generated over an object database schema. We use this language for implementing methods and calculations without side effects, and for assisting our algebraic query language. Within this language, our special subtyping order allows us to give a formalization of different kinds of binding modes in object programming including early and late binding. In fact, this language is inspired from category theory as a fragment of a the Kleisli category of the lifting monad. However, in this paper the category aspects of the model have not been discussed.

To summarize, our main contributions in this paper are the followings:

- to provide formal definitions for object-oriented database concepts namely, schema, inheritance, method, instance, null value, and query; and to give a formal solution to the problem of hierarchical consistency of a schema;
- to define an algebraic query languages which extend clearly the relational algebra and the nested relational algebra with null values, and which is enable to express SQL or OQL queries.
- to deal rigorously with null values; and
- to define a functional variable free language for defining calculation on objects and giving an operational and an algebraic semantics for this language, and to give a formalism for different binding modes.

The rest of this paper is organized as follows. In Section 2, we present the background material on the algebraic aspects of the relational model. This makes the paper self-contained and permits to better see the similarity of our object model with the relational style . Section 3 is the core of the paper. In this section we explores, in a formal way, the syntax and the semantics of our object model. In Section 4 we define an algebraic query language for the model. In Section 5 we introduce the syntax and the semantics of the kernel of a language serving to write code of methods and assisting the query language. Related works are discussed in Section 6 and we draw our conclusions and further works in Section 7. For the sake of clarity, we gather all examples in Appendix A and all proofs of Theorems and Facts in Appendix B.

2   Relational Databases Revisited

2.1   Relational Database Schema

We begin with three enumerable and pairwise disjoint sets REL, ATT, and DOM. We assume each of these sets is endowed with a syntactic equality that we denote by $=$. In practice an element of DOM represents a type name in a type system, or a synonym for such a type name. In the relational model this type system contains some predefined basic types only. That is type::= BASE and

$$\text{BASE ::= } \mathbf{int} \mid \mathbf{char} \mid \cdots \tag{1}$$

Definition 1 *A relational schema (over ATT, and DOM) is a finite partial function from ATT to DOM which is not the everywhere undefined function. A relational database schema (over REL, ATT, and DOM) consists of a non empty finite subset $\mathcal{R}$ of REL and for each $R \in \mathcal{R}$ a relational schema $\bar{R}$ (over ATT, and DOM) such that:*

$$\forall R, R' \in \mathcal{R}, \forall a \in \text{ATT } ((\bar{R}(a) = d) \wedge (\bar{R}'(a) = d')) \Rightarrow d = d'). \tag{2}$$

Condition 2 is called the "unique domain assumption", and R the name of $\bar{R}$. ∎

We will represent each relational schema of a relational database schema by its name, only. Indeed, names are distinct but many names may correspond to the same relational schema. For instance we will write $R(a)$ instead of $\bar{R}(a)$. A pair (a, d) in the graph of a relational schema R is called an *attribute* of R with name a and *domain* d, and is usually denoted by $a : d$. The unique domain assumption says: *in a relational database two attributes with the same name have the same domain.* Therefore, if $att(R)$ is the domain of definition of R and $\mathcal{U} = \bigcup_{R \in \mathcal{R}} att(R)$ then we can define a function $dom : \mathcal{U} \to \text{DOM}$ where $dom(a) = d$ means $a : d$.[1]

---

[1] In fact, condition 2 says that the relational schemas of a database schema are consistent partial functions and *dom* is their *join*, that is $dom_{|att(R)} = R$. But, we will not use this terminology because consistency has another meaning in database theory.

Note that att also defines a function $att : \mathcal{R} \to \mathcal{P}_f(\text{ATT}) \backslash \{\emptyset\}$. An alternative equivalent way to define a database schema is a triple $\mathcal{S} = (\mathcal{R}, att, dom)$ such that $\mathcal{R}$ is finite subset of REL and $att : \mathcal{R} \to \mathcal{P}_f(\text{ATT}) \backslash \{\emptyset\}$ and $dom : \mathcal{U} \to \text{DOM}$ are two functions, where $\mathcal{U} = \bigcup_{R \in \mathcal{R}} att(R)$. Then each element of $\mathcal{U}$ is called an attribute of $\mathcal{S}$. If R is a relational schema, $att(R) = \{a_1, \cdots, a_n\}$ and $dom(a_i) = d_i$ $(1 \leq i \leq n)$ then the declaration $R(a_1 : d_1, \cdots, a_n : d_n)$ defines entirely the relational schema R. Therefore, if $\mathcal{R} = \{R_1, \cdots, R_k\}$ then the database schema $\mathcal{S}$ can be defined by k declarations:

$$R_1(a_1^1 : d_1^1, \cdots, a_{n_1}^1 : d_{n_1}^1), \cdots, R_k(a_1^k : d_1^k, \cdots, a_{n_k}^k : d_{n_k}^k)$$

in which if $i \neq j$ then $R_i \neq R_j$ and if $a_i^j = a_{i'}^{j'}$ then $d_i^j = d_{i'}^{j'}$.

2.2 Relational Database Instance

Let $\mathcal{S} = (\mathcal{R}, att, dom)$ be a database schema over ATT and DOM. Let us call a *domain interpretation function* for $\mathcal{S}$, any function which associates with each domain name, occurring in $\mathcal{S}$, an enumerable set $[\![d]\!]$. The set $[\![d]\!]$ is called the *interpretation* of d. We assume that $[\![d]\!]$ is endowed with a strict order relation $<_d$ and an equality $=_d$. An element of $[\![d]\!]$ is called a *value* of type d. In practice, d represents a type name (or a synonym for a type name) and $[\![d]\!]$ is the associated concrete type, hence $[\![d]\!]$ is understood by d. For instance, in the standard database language SQL, the CREATE DOMAIN ... command defines a synonym of a type. Let $R(a_1 : d_1, \cdots, a_n : d_n)$ be a relational schema over ATT and DOM (not necessarily in $\mathcal{S}$) such that d_i is a domain name occurring in $\mathcal{S}$ for $i = 1, \cdots, n$. A function $t : \{a_1, \cdots, a_n\} \to \bigcup_{1 \leq i \leq n} [\![d_i]\!]$ is called a *tuple* of R, or a tuple over $\{a_1, \cdots, a_n\}$ if $t(a_i)$ is in $[\![d_i]\!]$ for $i = 1, \cdots, n$. In practice, tuples represent *data* of a real world. A finite set of tuples of R is called an *extension* of R. The set of all extensions of R is denoted $\mathcal{E}xt(R)$.

Definition 2 *A relational database instance over the relational database schema* $\mathcal{S}$ *is a pair* $\mathcal{I} = ([\![\,]\!], \alpha)$ *consisting of a domain interpretation function* $[\![\,]\!]$ *and a function* $\alpha : \mathcal{R} \to \bigcup_{R \in \mathcal{R}} \mathcal{E}xt(R)$ *such that* $\alpha(R) \in \mathcal{E}xt(R)$, *for any* $R \in \mathcal{R}$. ∎

We refer to this definition of instance as the *tuples-as-functions* approach.
By a *relational database* we mean a database relational schema accompanied with a relational instance of this schema (which may be the empty instance). In practice in a database $(\mathcal{S}, \mathcal{I})$ for any $R \in \mathcal{R}$ the triple $(R, att(R), \alpha(R))$ is represented as a table with name R, heading row $att(R)$, and other rows elements of $\alpha(R)$. The problem of defining a database instance is the same as defining a set of finite sets. In the relational database model, the database instance is "defined by extension": the instance is the set of stored tuples. However, this instance must satisfy some conditions called (*integrity*) *constraints*. Constraints impose restrictions on values allowed in an instance. In practice the constraints are specified within the schema and are checked automatically by the system

after any request of users for modification of the database instance. A database is said to be *consistent* if its instance satisfies its specified constraints. From a theoretical point of view many kinds of constraints have been proposed but in commercial systems only a few class of these constraints have been implemented. Within a *database language* one can specify a database schema and its constraints, and an instance of this schema. For example, in SQL a set of CREATE TABLE ... commands perform this task by creating a database schema and an empty instance, INSERT INTO ... VALUES ... commands add tuples into the database instance.

2.3   Relational Algebra

Querying a database consists of extracting data from the database. In fact the extracted data corresponds to the answer of a query on the real world objects, represented by the database. A relational query language allows us to specify a query on the real world objects in the form of a relational schema, and to provide its answer in the form of an extension of this relational schema. The schema of the query as well as its answer are obtained by combining relational schemas and relational instances of the database. In commercial languages users specify the schema of a question, and gives some information for finding its answer. The system finds automatically the answer to the question as an extension of that schema. In other words, evaluation process is transparent to users. For example, the SELECT ... FROM ... WHERE ... command of SQL performs this task. It is customary to call *query* the specification of a question and to call *evaluation of the query* the computation of its answer. The kernel of each query language for the relational model, is the implementation of the relational algebra. Relational algebra seen as a query language consists of specifying and evaluating queries as algebraic expressions over the database. In the sequel we will consider a database with the schema $\mathcal{S} = (\mathcal{R}, att, dom)$ and a database instance $([\![]\!], \alpha)$ over $\mathcal{S}$. We will denote by the same symbol, say θ, any of the two binary relations $<_d$ and $=_d$ of any $[\![d]\!]$.

Query Specification:

For specifying a query q over $\mathcal{S}$ we have to define the algebraic expression of q and the set of attributes of q. In other words, we have to define the set $\mathcal{Q}$ of queries and a function $att^* : \mathcal{Q} \to \mathcal{P}_f(\text{ATT} \backslash \{\emptyset\})$. To this end, we introduce and recall some notations. An *elementary condition* over $\mathcal{S}$ is a declaration of the form $a\theta v$, $v\theta a$, or $a\theta b$ where a and b are attributes of $\mathcal{S}$ and $v \in [\![dom(a)]\!]$. A *condition* is a first order formula whose atoms are elementary conditions. A condition *cond* is said to be a condition *over* a relational schema R (not necessary in $\mathcal{S}$) if attributes occurring in *cond* are in $att(R)$. Let us see again R as a partial function $R : \text{ATT} \to \text{DOM}$ and consider $b \in att(R)$ and $c \in \text{ATT} \backslash att(R)$ such that if $c \in \mathcal{U}$ then $dom(c) = dom(b)$. Now, define a new relational schema as follows:

$$\Re_{b \mapsto c}(R)(a) = \begin{cases} R(b) & \text{if } a = c \\ \text{undefined} & \text{if } a = b \\ R(a) & \text{otherwise} \end{cases}$$

We say $\Re_{b \mapsto c}(R)$ is obtained from R by *renaming* attribute b to c. More generally, let $L_1 = [b_1, \cdots, b_m]$ and $L_2 = [c_1, \cdots, c_m]$ be two lists of attribute names such that elements of L_1 (of L_2) are distinct from each other, $b_i \in att(R)$, $c_i \notin [b_i, \cdots, b_m]$ and if $c_i \in \mathcal{U}$ then $dom(c_i) = dom(b_i)$ ($1 \leq i \leq m$). It is not difficult to see that the expression $\Re_{b_1 \mapsto c_1}(\cdots \Re_{b_m \mapsto c_m}(R) \cdots)$ is a well defined expression. We denote this expression by $\Re_{L_1 \mapsto L_2}(R)$, and we say (L_1, L_2) is a *renaming pair* for R. Note that the renaming operation doesn't introduce new domains, but new attribute names. Indeed, $att(\Re_{L_1 \mapsto L_2}(R)) = (att(R) \backslash \{b_1, \cdots, b_m\}) \cup \{c_1, \cdots, c_m\}$. Let R and R' be two relational schemas (not necessarily in $\mathcal{S}$) such that any $a \in att(R) \cap att(R')$, has the same domain in R and in R'. Then we can define a new relational schema $R \bowtie R'$ such that $att(R \bowtie R') = att(R) \cup att(R')$, and domains in $R \bowtie R'$ are the same as in R and R'.

Definition 3 *The set $\mathcal{Q}$ of queries over $\mathcal{S}$ and the function $att^* : \mathcal{Q} \rightarrow \mathcal{P}_f(\textbf{ATT} \backslash \{\emptyset\}$ are defined recursively as follows:*

- *if $R \in \mathcal{R}$ then $R \in \mathcal{Q}$ and $att^*(R) = att(R)$,*
- *if $q \in \mathcal{Q}$ and cond is a condition over q then $\sigma_{cond}(q) \in \mathcal{Q}$ and $att^*(\sigma_{cond}(q)) = att^*(q)$,*
- *if $q \in \mathcal{Q}$ and $X \subseteq att^*(q)$ then $\prod_X(q) \in \mathcal{Q}$ and $att^*(\prod_X(q)) = X$,*
- *if $q \in \mathcal{Q}$ and (L_1, L_2) is a renaming pair for q then $\Re_{L_1 \mapsto L_2}(q) \in \mathcal{Q}$ and $att^*(\Re_{L_1 \mapsto L_2}(q)) = att(\Re_{L_1 \mapsto L_2}(q))$,*
- *if $q_1 \in \mathcal{Q}$ and $q_2 \in \mathcal{Q}$ are two queries then $q_1 \bowtie q_2 \in \mathcal{Q}$ and $att^*(q_1 \bowtie q_2) = att^*(q_1) \cup att^*(q_2)$,*
- *if $q_1 \in \mathcal{Q}$ and $q_2 \in \mathcal{Q}$ and $att^*(q_1) = att^*(q_2) = A$ then $q_1 \cup q_2$ and $q_1 - q_2$ are in $\mathcal{Q}$ and $att^*(q_1 \cup q_2) = att^*(q_1 - q_2) = A$,*
- *the only elements of $\mathcal{Q}$ are those obtained as above.* ∎

Note that the above definition uses the schema $\mathcal{S}$ and the domain interpretation function $[\![ ]\!]$, only. Operations σ, $\prod$, $\Re$, $\bowtie$, $\cup$ and - are called `selection`, `projection`, `renaming`, `join`, `union` and `difference`, respectively. They form a partial (universal) algebra called the *relational algebra*.

Query Evaluation:

Evaluating a query q over a database instance $\mathcal{I}$ consists of extracting a particular extension of q from $\mathcal{I}$. By extension of q we mean a finite set of tuples over $att^*(q)$. In other words, the evaluation process consists of defining a convenient function $\alpha^* : \mathcal{Q} \rightarrow \bigcup_{q \in \mathcal{Q}} \mathcal{E}xt(q)$ such that $\alpha^*(q) \in \mathcal{E}xt(q)$. In order to define α^* we recall and introduce some notations. Let $R(a_1 : d_1, \cdots, a_n : d_n)$ be a relational schema over **ATT** and **DOM** (not necessarily in $\mathcal{S}$) such that d_i is a domain name occurring in $\mathcal{S}$ for $i = 1, \cdots, n$. Let $cond$ be a condition and t a tuple over R. Let $b_1, \cdots, b_m$ be attributes of R occurring in $cond$, and $[t(b_1)/b_1, \cdots, t(b_m)/b_m]$ the boolean expression obtained from $cond$ by $t(b_i)$ substituted for b_i. We say t *satisfies* $cond$, denoted $t \models cond$, if this boolean expression is evaluated to true. If r is an extension of R the set $\{t \in r \mid t \models cond\}$ is denoted $\sigma_{cond}(r)$, and the set $\{t_{\mid X} \mid t \in r\}$ is denoted $\prod_X(r)$, where $t_{\mid X}$ means the restriction of t on X. Let R' be another relational schema and t' a tuple of R' such that $t(a) = t'(a)$ for every $a \in att(R) \cap att(R')$. Then we denote by $t \bowtie t'$ the tuple over $R \bowtie R'$ such that $(t \bowtie t')_{\mid att(R)} = t$ and $(t \bowtie t')_{\mid att(R')} = t'$. If r is an extension of R and r' an extension of R' then $r \bowtie r' = \{t \bowtie t' \mid t \in r, t' \in r' \text{ and } t \bowtie t' \text{ is defined}\}$. Moreover, we suppose $r \bowtie r'$ is the empty set whenever there is no $t \in r$ and no $t' \in r'$ such that $t \bowtie t'$ is defined.

Definition 4 *The evaluation function* $\alpha^* : \mathcal{Q} \to \bigcup_{q \in \mathcal{Q}} \mathcal{E}xt(q)$ *is defined recursively as follows:*

$\forall R \in \mathcal{R}$, $\alpha^*(R) = \alpha(R)$,

$\alpha^*(\sigma_{cond}(q)) = \sigma_{cond}(\alpha^*(q))$, $\qquad \alpha^*(\prod_X(q)) = \prod_X(\alpha^*(q))$,

$\alpha^*(\Re_{L_1 \mapsto L_2}(q)) = \alpha^*(q)$, $\qquad \alpha^*(q_1 \bowtie q_2) = \alpha^*(q_1) \bowtie \alpha^*(q_2)$,

$\alpha^*(q_1 \cup q_2) = \alpha^*(q_1) \cup \alpha^*(q_2)$, $\qquad \alpha^*(q_1 - q_2) = \alpha^*(q_1) - \alpha^*(q_2)$. ∎

Since the database is finite, the evaluation of any query provides a finite set of tuples.

2.4   Relational Databases with Null Values

The value of an attribute in a given instance may be undefined. Many kinds of undefinedness have been pointed out in the literature including values unknown in the current instance, values which can not exist in the context of application, and so on. An undefined value is called *null value* in database jargon and is denoted null. We consider a unique kind of null value, denoted by $\bot$. We see $\bot$ as a symbol which doesn't represent an element in any domain. To deal with the null value we have to consider a tuple over a schema $\{a_1 : d_1, \cdots, a_n : d_n\}$ as a function $t : \{a_1, \cdots, a_n\} \to \bigcup_{1 \leq i \leq n} [\![d_i]\!]_\bot$ where $[\![d_i]\!]_\bot = [\![d_i]\!] + \{\bot\}$ and $t(a) = \bot$ means $t(a)$ is undefined. Now, evaluating conditions needs a three valued logic with truth values true, false, and undef. In this logic $\bot \theta v$ is evaluated to false and $\bot = \bot$ to true, while $\bot < v$, $v < \bot$ and $\bot < \bot$ are evaluated to undef. The logical connectors are extended to undef as follows:

$\wedge$	true	false	undef
undef	undef	false	undef

$\vee$	true	false	undef
undef	true	undef	undef

$\neg$	undef
undef	undef

Now, we extend our elementary conditions with the declaration $A = \bot$ (which corresponds to the SQL command A is null. The above consideration seems to be enough for treating null values at least in the style of SQL. There are various other approaches for null values in the literature [26].

In the rest of the paper we will extend this style of relational model to the object models by adding the three basic concepts of objects namely inheritance, methods, and object identity.

3 The Object Data Model

3.1 A Type System for Object Modeling

We begin with two enumerable disjoint non empty sets BASE and CLASS. In practice, BASE and CLASS are sets of names representing *basic types* and *class names*, respectively. Thus, we assume each of these two sets is endowed with a syntactic equality denoted $=$. Next, we define a type system over BASE and CLASS in which types are built recursively by the following grammar:

$$\text{type} ::= \text{BASE} \mid \text{CLASS} \mid \text{type} \otimes \text{type} \mid \text{coll type}. \tag{3}$$

The constructor *coll* is a collection constructor like *set*, *bag*, *list*, etc (i.e. *coll* ::= *set* | *bag* | *list* | $\cdots$). Herein, however, we restrict our study to *set* (see [21], [25] for more detailed formalization on collection types). The constructor $\otimes$ is a product but its semantics will not be the cartesian product semantics. Elements of type are called *type expressions* or *object-types* (or simply *types*). We use parentheses only when there is risk of ambiguity and in order to avoid proliferation of parentheses we use $t_1 \otimes t_2 \otimes \cdots \otimes t_n$ as an abbreviation for $t_1 \otimes (t_2 \otimes (\ldots \otimes t_n) \ldots)$. In the sequel, for any finite subset C of CLASS we denote by $\mathcal{T}_C$ the set of type expressions generated by BASE and C. That is,

$$\mathcal{T}_C ::= \text{BASE} \mid C \mid \mathcal{T}_C \otimes \mathcal{T}_C \mid set\ \mathcal{T}_C. \tag{4}$$

Then we denote the empty sequence of types by unit, the set of all sequences of $\mathcal{T}_C$ excluding unit by $\mathcal{T}_C^+$, the set $\mathcal{T}_C^+ \cup \{\text{unit}\}$ by $\mathcal{T}_C^*$, and the set $\mathcal{T}_C \cup \{unit\}$ by $\mathcal{T}_C^u$. Thus, a sequence of types is determined in a unique way as $r = t_1 \cdots t_n$, where $n > 0$ and $t_i \in \mathcal{T}_C$ for all i ($1 \le i \le n$), or $r = unit$ ($n = 0$). We omit the subscript C whenever there is no possibility of confusion. The syntactic equality of BASE and CLASS can be extended to $\mathcal{T}_C$, $\mathcal{T}_C^+$, $\mathcal{T}_C^*$ and $\mathcal{T}_C^u$ in an obvious way.

3.2 Object Database Schema

We consider two enumerable non-empty sets ATT (attribute names) and METH (method names). We assume that ATT and METH are disjoint sets, endowed with a syntactic equality $=$, and have no common elements with BASE and CLASS.

Definition 5 *We say* $\mathcal{S} = (C, isa, att, meth)$ *is an object-oriented database schema or an* object schema *(or a schema for short) if:*

- C *is a finite non-empty subset of* CLASS,
- *isa is a binary acyclic relation over* C, *and*
- $att : C \times \text{ATT} \rightarrow \mathcal{T}_C$ *and* $meth : C \times \text{METH} \times \mathcal{T}_C^* \rightarrow \mathcal{T}_C^u$ *are two partial functions*

such that for every c in C at least one of the three following sets isa(c), att(c) or meth(c) is non-empty (i.e. $att(c) \cup isa(c) \cup meth(c) \neq \emptyset$).

$$isa(c) \quad = \{c' \mid c \ isa \ c'\}$$
$$att(c) \quad = \{(a,t) \mid att(c,a) = t\}$$
$$meth(c) = \{(m,r,t) \mid meth(c,m,r) = t\}. \quad \blacksquare$$

The pair (C, att) is said to be the *structural schema*, the pair $(C, meth)$ the *behavioral schema*, and the relation *isa* the *inheritance hierarchy*. For each class name c in C, $(c, isa(c), att(c), meth(c))$ is called a *class* of S with name c. Definition 5 implies that classes of a schema have distinct names. Thus, in a schema, a class can be recognized by its name, and we shall do so in the sequel.

According to the usual notation in the object-oriented paradigm, we read *c isa c'* as *c inherits c'*. The acyclic condition implies that *isa* is not symmetric. We say inheritance is *multiple* if there is at least one class c which inherits two classes c_1 and c_2 such that c_1 and c_2 are not comparable for the inheritance relation *isa*, otherwise inheritance is said to be simple. Our definition of schema does not impose any restrictions on inheritance. Simple inheritance as well as multiple inheritance are allowed. The functionality of *att* in Definition 5 says that : *overloading attribute names in a class is not allowed.*

Each (a,t) in $att(c)$ is displayed as $a : t$ and is called an *attribute* of c with name a and *type t*. The functionality of *att* in Definition 5 means that *"overloading of attributes is not allowed within a class"*. However, this does not prevent two attributes to be overloaded in two distinct classes. In other words, the unique domain assumption of the relational model (condition 2) is satisfied within a class, only. Therefore, an attribute (a,t) of a class c should be seen as (c, a, t) in the whole schema. However, when no confusion is possible we identify an attribute by its name.

Each (m, r, t) in $meth(c)$ is displayed as $m : t_1 t_2 \ldots t_k \rightarrow t$ (where $r = t_1 t_2 \ldots t_k$) and as $m : \rightarrow t$ (where $r = unit$), and is called a method of c with name m, *rank* (or *type parameter*) r, and *type t*. The pair (r, t) is the *profile* of the method. Roughly speaking, methods represents parameterized computations or updates over objects. Note that the rank and/or the type of a method can be the type *unit*. A method whose rank is *unit* has no parameters and a method whose type is *unit* has no result. The latter represents a computation with side-effects or an output command. The functionality of *meth* in Definition 5 says that:

> *"two methods in a class can have the same name providing that they do not have the same rank".*

However, this does not prevent two methods in distinct classes having the same rank . Thus, overloading methods is a more sophisticated problem.

The last condition of Definition 5, namely $att(c) \cup isa(c) \cup meth(c) \neq \emptyset$, says that a class cannot be specified only by its name. It must have at least one attribute or one method, otherwise it must inherit another class.

The above notation agrees with the semantics that we intend to introduce later on. In this semantics, invoking a method on an object will correspond to calling a function or a procedure, a class will be viewed as a container of objects, and an object will be recognized by its state and its behavior. The inheritance will permit objects to inherit structure and behavior from other objects.**Syntactic Sugar**

Our definition of schema and class is similar to the class declaration of most object-oriented data models. In these models the declaration of a class looks like the following, up to some syntactic conventions.

> **class** c
>> **inherits** $c_1, \ldots, c_m$
>> **attributes**
>>> $a_1 : t_1, \cdots, a_n : t_n$
>> **methods**
>>> $m_1 : \quad t_1^1 \ldots t_{k_1}^1 \to t^1,$
>>> $\vdots$
>>> $m_p : \quad t_1^p \ldots t_{k_p}^p \to t^p;$

In our setting this corresponds to:

$$isa(c) = \{c_1, \ldots, c_m\}$$
$$att(c) = \{(a_1, t_1), \ldots, (a_n, t_n)\}$$
$$meth(c) = \{(m_1, t_1^1 \ldots t_{k_1}^1, t^1), \ldots, (m_p, t_1^p \ldots t_{k_p}^p, t^k)\}.$$

Therefore, a finite set of such declarations forms a schema iff the resulting binary relation isa is acyclic, att and $meth$ are functional and the last condition of Definition 5 is satisfied (see Appendix A for a concrete example).

3.3   Hierarchical Consistency of an Object Database Schema

Subtyping

Let us denote by $\leq_{isa}$ the transitive and reflexive closure of isa. Since $\leq_{isa}$ is a partial order we write $c <_{isa} c'$ for $(c \leq_{isa} c') \wedge (c \neq c')$ and we read $c <_{isa} c'$ as c *is a subclass of* c' or c' *is a superclass of* c. The set of all super classes of c is denoted $super(c)$ and we write $super^*(c) = super(c) \cup \{c\}$. Sets $sub(c)$ and $sub^*(c)$ are defied similarly for subclasses.

Fact 1 *If inheritance is simple then for any c and any property p the set $\{c \mid p(c) \wedge (c \in super^*(c))\}$ has a minimum, whenever it is not empty.*

Indeed, in a simple inheritance $super^*(c)$ is a chain for $\leq_{isa}$. ∎

The following extends the partial order $\leq_{isa}$ to a partial order $\preceq_{\mathcal{T}_C}$ on all types.

Definition 6 *The subtyping relation $\preceq_{\mathcal{T}_C}$ on $\mathcal{T}_C$ is defined recursively as follows:*

1. *if b is a basic type then $b \preceq_{\mathcal{T}_C} b$;*
2. *if $c \leq_{isa} c'$ then $c \preceq_{\mathcal{T}_C} c'$;*
3. *if $t \preceq_{\mathcal{T}_C} t'$ then $(set\ t) \preceq_{\mathcal{T}_C} (set\ t')$;*
4. *if $t_1 \preceq_{\mathcal{T}_C} t_1'$ and $t_2 \preceq_{\mathcal{T}_C} t_2'$ then $(t_1 \otimes t_2) \preceq_{\mathcal{T}_C} t_1' \otimes t_2'$;*
5. *$t \preceq_{\mathcal{T}_C} t'$ is defined only by rules 1-4.* ∎

One can prove by structural recursion that $\preceq_{\mathcal{T}_C}$ is a partial order. As usual $t \prec t'$ abbreviates $(t \preceq t') \wedge (t \neq t')$. We read $t \prec t'$ as " t is a *subtype* of t' ". It is clear that the above subtype ordering is deduced only from inheritance. Multiple and simple subtyping can be defined in the same way as for inheritance. In practice, basic types may also be endowed with a predefined and simple subtyping relation $\leq_{\text{BASE}}$. To encompass such situations in our formalism, we can replace clause 1 of the above definition by the following:

1. *if b_1 and b_2 are basic types and $b_1 \leq_{\text{BASE}} b_2$, then $b_1 \preceq_{\mathcal{T}_C} b_2$.*

However, for the rest of this paper we assume that $\preceq_{\mathcal{T}_C}$ is given by Definition 6.

Fact 2 *Let $\mathcal{S} = (C, isa, att, meth)$ be a schema. The subtype ordering $\preceq_{\mathcal{T}_C}$ is not necessarily simple even if the inheritance relation isa is simple.* ∎

The partial order $\preceq_{\mathcal{T}_C}$ can be easily extended to a partial order $\preceq_{\mathcal{T}_C}^*$ on $\mathcal{T}_C^*$ by setting $unit \preceq_{\mathcal{T}_C}^* unit$, and $t_1 \cdots t_n \preceq_{\mathcal{T}_C}^* t_1' \cdots t_m'$ iff $n = m$ and $t_i \preceq_{\mathcal{T}_C} t_i'$ for all i, $1 \leq i \leq n$. The restriction of $\preceq_{\mathcal{T}_C}^*$ to $\mathcal{T}_C^u$ is denoted by $\preceq_{\mathcal{T}_C}^u$. In the sequel we will refer to any of these order relations as subtype relations and we will omit the subscripts when they are understood from the context. Note that our subtyping relation is generated by *isa* and is different from the classical subtyping in object programming languages. For example, in our setting $t \otimes t'$ is not a subtype of t while in the classical subtyping relation $t \times t'$ is a subtype of t. Another important consequence of our subtyping is the following fact which can be proved easily by structural recursion.

Fact 3 *For any given type t, the set $super(t)$ of super-types of t, and the set $sub(t)$ of subtypes of t are finite.* ∎

Overloading and Overriding Problems

An important concept in the object paradigm is *overriding* which is a particular case of our overloading. Roughly speaking, overriding consists of hiding an attribute or a method of a class in a subclass by redefining it.

Definition 7 (Overriding) *We say that an attribute $a : t'$ of a class c' is overridden in a class c if c is a subclass of c' and there is an attribute $a : t$ in c. We say that a method $m : r' \to t'$ of a class c' is overridden in a class c if $c \leq_{isa} c'$ and there is a method $m : r \to t$ in c such that $r \preceq_{\mathcal{T}_C}^* r'$.* ∎

A method can be overridden more than once in a class. Thus, we have to provide techniques for determining which one really hides the corresponding method in the superclass. Our overriding concept is similar, but more general, to that of the object-oriented model O2 [20]. In O2 it is not allowed to overload an overridden method. In the object programming language JAVA [27] overriding is strong in the sense that the two methods, in the class and in the subclass, must have the same profile. The problem of overriding is a delicate and important problem in the object-oriented paradigm. Indeed, overridden methods can cause inconsistency during binding and Definition 5 can lead to such inconsistency. To avoid such inconsistecies some conditions have to be satisfied.

Definition 8 (covariance) *The database schema $\mathcal{S}$ is said to satisfy the covariance condition iff*

1. *for every attribute $a : t'$ in a class c', if $a : t'$ is overridden as $a : t$ in a class c then $t \preceq t'$, and*
2. *for every method $m : r' \to t'$ in a class c', if $m : r' \to t'$ is overridden in the class c as $m : r \to t$ then $t \preceq^u_{\mathcal{T}_C} t'$.* ∎

Because of overloading, an attribute name is related to several types and a method name to several profiles. The covariance conditions say how these types or profiles are related with subtype ordering. Covariance can be checked automatically following Fact 3.

Let us define a partial order on the domain of partial function $att : C \times \text{ATT} \to \mathcal{T}_C$, and a partial order on the domain of the partial function $meth : C \times \text{METH} \times \mathcal{T}_C^* \to \mathcal{T}_C^u$ as follows:

$$(c, a) \leq_{C \times \text{ATT}} (c', a') \iff c \leq_{isa} c' \text{ and } a = a'.$$
$$(c, m, r) \leq_{C \times \text{METH} \times \mathcal{T}_C^*} (c', m', r') \iff c \leq_{isa} c' \text{ and } m = m' \text{ and } r \preceq^* r'.$$

Obviously, Definition 8 implies the following :

Fact 4 *A schema $\mathcal{S} = (C, isa, att, meth)$ satisfies the covariance condition iff the partial functions $att : C \times \mathcal{A} \to \mathcal{T}_C$ and $meth : C \times \text{METH} \times \mathcal{T}_C^* \to \mathcal{T}_C^u$ are monotonic for the above ordered sets.* ∎

Derived Attributes and Derived Methods

Our definition of database schema allows us to use an attribute (or method) name explicitly in several classes with probably different types (or different profiles) in each class. The problem arises when an explicit attribute (or method) of a superclass must be seen implicitly as an attribute (or a method) present in a subclass. In other words, what are conditions allowing to derive attributes and methods unambiguously. In what follows we shall answer this question formally. To this end, we consider the following partial order over the set $C \times \mathcal{T}_C^*$

$$(c, r) \preceq_{C \times \mathcal{T}_C^*} (c', r') \iff c \leq_{isa} c' \text{ and } r \preceq_{\mathcal{T}_C^*} r'$$

and then we define

$super^*(c, a) = \{c' \mid (c' \in super^*(c)) \land (\exists t \in \mathcal{T}_C \ att(c', a) = t\}$, and
$super^*(c, m, r) = \{(c', r') \mid (c, r) \preceq_{C \times \mathcal{T}_C^*} (c', r') \land (\exists t \in \mathcal{T}_C^u \ meth(c', m, r') = t\}$.

An element of $super^*(c, a)$ is either c or a superclass of c in which a is the name of an attribute. An element of $super^*(c, m, r)$ is either the pair (c, r) or a pair (c', r') formed by a class name $c' \in super^*(c)$ in which m is the name of a method with argument types r' and $r \preceq_{\mathcal{T}_C^*} r'$.

Definition 9 (minimum condition) *Let S be a schema. We say att satisfies the minimum condition iff for each class c and each attribute name a occurring in S, the subset super*(c, a) has a minimum in the ordered set $(C, \leq_{isa})$, whenever this subset is not empty. We say meth satisfies the minimum condition if for each class c, each method name m and each sequence r of types occurring in S, the subset super*(c, m, r) has a minimum in the ordered set $(C \times \mathcal{T}_C^*, \leq_{C \times \mathcal{T}_C^*})$, whenever this subset is not empty. The schema S satisfies the minimum condition if att and meth satisfy this condition.* ∎

Since $super^*(c, a)$ and $super^*(c, m, r)$ are finite sets, the minimum conditions can be checked automatically. When these minimums exist we write $ResA(c, a) = min(super^*(c, a))$ and $ResM(c, m, r) = min(super^*(c, m, r))$ and we call them the *resolution of a in c* and the *resolution of m in c with respect to r*, respectively. The resolution technique allows us to compute the type of a derived attribute or method. Indeed, if $ResA(c, a) = c'$ then there is an attribute $a : t'$ in c' and $c \leq_{\mathcal{T}_C} c'$. If $c' = c$ then $a : t'$ is explicitly defined in c too, that is $att(c, a) = t'$. Otherwise, $att(c, a)$ is not defined explicitly, but $a : t'$ may be seen as an implicit (hidden or derived) attribute of c. We can extend att to (c, a) by defining $\overline{att}(c, a) = t'$. In a similar way, if $ResM(c, m, r) = (c', r')$ then there is a method $m : r' \to t'$ in c'. Then, either $(c, r) = (c', r')$ and $m : r' \to t'$ is an explicit method in c that is $meth(c, m, r') = t'$, or $m : r' \to t'$ is a derived method in c and we can define $\overline{meth}(c, m, r') = t'$. Therefore, when the schema satisfies the minimum condition the following rules extend the functions att and $meth$ to all derived attributes and derived methods, respectively:

$$\frac{ResA(a, c) = c' \quad att(c', a) = t'}{\overline{att}(c, a) = t'} \tag{5}$$

$$\frac{ResM(m, c, r) = (c', r') \quad meth(c', m, r') = t'}{\overline{meth}(c, m, r) = t'} \tag{6}$$

That is, $\overline{att} : C \times \mathtt{ATT} \to \mathcal{T}_C$ and $\overline{meth} : C \times \mathtt{METH} \times \mathcal{T}_C^* \to \mathcal{T}_C^u$ are two partial functions more defined than att and $meth$, respectively.

Fact 5 *If inheritance is simple then att satisfies the minimum condition. However, meth may not satisfy the minimum condition even if inheritance is simple.* ∎

Definition 10 *We say a schema $S = (C, isa, att, meth)$ is hierarchically consistent if it satisfies the covariant and the minimum condition.* ∎

Hierarchical consistency says how the inheritance relation of a schema has to operate on attributes and methods. In other words, it says how an attribute or method of a class can be considered as an attribute or as a method in its subclasses unambiguously. Checking inheritance can be done automatically.

Theorem 1 *If $S = (C, isa, att, meth)$ is a hierarchically consistent schema then $\overline{S} = (C, isa, \overline{att}, \overline{meth})$ is also a hierarchically consistent schema that we call the closure of S. Moreover, $\overline{(\overline{S})} = \overline{S}$.* ∎

3.4   Object-Oriented Databases

Semantics of Types

As we have seen earlier, a type is either the *unit* type or an expression built-up from basic types and class names using the constructors $\otimes$ and *set*. From now on, we assume `bool` is the name of a basic type. We will interpret each type t by a set $[\![t]\!]$. To this end, we interpret each basic type t by an enumerable set $[\![t]\!]$ seen as the concrete type of t. Concrete types are assumed to be pairwise disjoint. We consider a special non-empty enumerable set `Oid` and a special symbol $\bot$. We assume that `Oid` is disjoint from all other basic concrete types, and the symbol $\bot$ denotes neither an element of a basic type nor an element of `Oid`.[2] We will denote the usual cartesian product by $\times$ and the usual cartesian co-product (i.e. disjoint union) by $+$. To simplify the description, we define $[\![t]\!]$ and an intermediate set $[\![t]\!]_\bot$ recursively as follows:

- $[\![unit]\!] = \{\bot\}$ and $[\![bool]\!] = [\![unit]\!] + [\![unit]\!]$.
- For any basic type b, $[\![b]\!]_\bot = [\![b]\!] + [\![unit]\!]$, and $[\![unit]\!]_\bot = [\![unit]\!] + [\![unit]\!]$.
- For any class name c, $[\![c]\!] = $ `Oid` and $[\![c]\!]_\bot = [\![c]\!] + [\![unit]\!]$.
-

$$[\![t_1 \otimes \cdots \otimes t_n]\!]_\bot = [\![t_1]\!]_\bot \times \cdots \times [\![t_n]\!]_\bot, \ n \geq 1 \tag{7}$$

-

$$[\![t_1 \otimes \cdots \otimes t_n]\!] = [\![t_1 \otimes \cdots \otimes t_n]\!]_\bot \backslash [\![unit]\!]^n, \ n \geq 1 \tag{8}$$

-

$$[\![set\ t]\!] = \mathcal{P}_f([\![t]\!]) \ \text{ and } \ [\![set\ t]\!]_\bot = [\![set\ t]\!] + [\![unit]\!] \tag{9}$$

where $[\![unit]\!]^n = \overbrace{(\bot, \cdots, \bot)}^{n \text{ times}}$, and $\mathcal{P}_f([\![t]\!])$ denotes the set of finite subsets of $[\![t]\!]$. As usual $v : t$ means v is a *value* of type t. Values of concrete types are called *observable values*, those of `Oid` *non-observable values* or *object identities*, and $\bot$ is called the *null value*. Later on we shall see another role of the symbol $\bot$ which will express the undefinedness of functions. It is clear that $[\![bool]\!]$ has two elements that we denote by `true` and `false`. It is important to note that the semantics of $\otimes$ is not the usual cartesian product semantics. For example, $[\![t_1 \otimes t_2]\!] = [\![t_1]\!] \times [\![t_2]\!] + [\![t_1]\!] \times [\![unit]\!] + [\![unit]\!] \times [\![t_2]\!]$.[3] As a result, if $v = (v_1, \cdots, v_n) \in [\![t_1 \otimes \cdots t_n]\!]$ then only some (but not all) components of v may be $\bot$. This implies that the pair (v_1, v_2) may have several types. For example, $(\bot, 1)$ is a value of type *bool* $\otimes$ *int* as well as a value of type *int* $\otimes$ *int*. Since $[\![unit]\!]^n \simeq [\![unit]\!]$ ($\simeq$ means isomorphic to) the above definition implies that:

---

[2] The assumption that all these sets are pairwise disjoint is not necessary if we ensure that they have distinct names. In fact a concrete type is accompanied with its name.

[3] Nevertheless, $\otimes$ is a product in the sense of category theory, a subject which is deliberately not discussed in this paper.

$$[\![t]\!]_\perp = [\![t]\!] + [\![unit]\!] \quad , \quad [\![t]\!] = [\![t']\!] \quad \text{iff} \quad [\![t]\!]_\perp = [\![t']\!]_\perp$$
$$[\![t_1 \otimes t_2]\!]_\perp \simeq [\![t_1]\!]_\perp \times [\![t_2]\!]_\perp.$$

We will use this isomorphism, especially when we compare a value v with $\perp$.

Semantics of Inheritance and Subtyping

An object identity has to identify a real world object. However, at the very outset, it cannot determine the nature of the real world object it represents. Indeed, all classes are interpreted as Oid, so even if two classes are related by inheritance, they have the same semantics. Moreover, the subtyping relation $\preceq_{\mathcal{T}_C}$ being induced only by the inheritance relation, we have that:

If $t_1 \preceq t_2$ then $[\![t_1]\!] = [\![t_2]\!]$, for any types t_1 and t_2.

This is a rather unpleasant result, and proves that more information is needed for associating types and values. In object databases and in object programming languages objects are assigned to classes (for instance by the procedure *new*). These kinds of assignments reduce the possible types of a value.

Definition 11 *Let C be a finite set of class names. An object classification function over C is any function $\gamma : C \to \mathcal{P}_f(\text{Oid})$ such that $\gamma(c) \cap \gamma(c') = \emptyset$ when $c \neq c'$. An object (identity) o is said to be γ-classified in c if $o \in \gamma(c)$.* ∎

With this definition, there might be classes in which no objects are classified and there might be objects which are not classified in any class. Clearly, classification of objects is independent of inheritance and subtyping. The operation new of most object-oriented systems is similar to our classification function, that is, $o := \text{new}(c)$ classifies a new o in c.

In a hierarchical system it is desirable to access a value of a class even if that value is not classified in that class, but in one of its subclasses. We say an object o is γ-*identifiable* in a class c iff o is γ-classified in one of its subclasses. γ-identification defines a function $[\![]\!]^\gamma : C \to \mathcal{P}_f(\text{Oid})$ by $[\![c]\!]^\gamma = \bigcup_{c' \leq_{isa} c} \gamma(c')$.

Denoting the set of all values by $\text{Val} = \bigcup_{t \in \mathcal{T}_C^u} [\![t]\!]$, this function can be naturally extended as a function $[\![]\!]^\gamma : \mathcal{T}_C^u \to \mathcal{P}_f(\text{Val})$ as follows:

- $[\![t]\!]^\gamma = [\![t]\!]$, if $t \in \text{BASE}$ or $t = unit$;
- $[\![(set\ t)]\!]^\gamma = \mathcal{P}_f([\![(t)]\!]^\gamma)$, for every type $t \in \mathcal{T}_C$, and
- $[\![t_1 \otimes \cdots \otimes t_n]\!]^\gamma = \{(v_1, \cdots, v_n) \mid (v_1, \cdots, v_n) \in [\![t_1 \otimes \cdots \otimes t_n]\!] \wedge$
  $\forall i \geq 1 \ \ v_i \in [\![t_i]\!]_\perp^\gamma\}.$

We see $[\![t]\!]^\gamma$ as *the semantics of the type t with respect to γ.*

Theorem 2 *For any classification function γ and all types t and t', if $t \preceq_{\mathcal{T}_C^u} t'$ then $[\![t]\!]^\gamma \subseteq [\![t']\!]^\gamma$.* ∎

This Theorem states that:

> With respect to a classification function the semantics of the subtyping relation (and in particular the semantics of the inheritance relation) is the set inclusion.

Nevertheless, even with a classification function γ, a value which is γ-identifiable in a type t could be γ-identifiable in some other types too. The question arises as to whether we could push one of these types ahead. This is the typical problem of object databases and object programming in which the static, declared type of a variable may be different from its dynamic, run-time type. The latter must be a subtype of the former, a property known as type-safety. In our approach, the subtype order is generated from inheritance. The dynamic type of a value is the most specific type in which that value is γ-identifiabe. The algorithm for computing *the most specific type* is given formally in the following definition.

Definition 12 *Let $v : t$ and $v \in [\![t]\!]^\gamma$. The most specific type of v with respect to t, denoted $mst(v : t)$, is defined recursively as follows:*

- *$mst(v : b) = b$ if b is a basic type or b is unit,*
- *$mst(v : t) = c$ if t is a class name and c is the name of the class in which v is γ-classified,*
- *$mst((v_1, \cdots, v_n) : t_1 \otimes \cdots \otimes t_n) = t'_1 \otimes \cdots \otimes t'_n$ $(n \geq 2)$, where*
$$t'_i = \begin{cases} t_i & \text{if } v_i = \bot \\ mst(v_i : t_i) & \text{otherwise} \end{cases}$$
- *$mst(v : set\ t) = \begin{cases} lub\{mst(v_i : t) \mid 1 \leq i \leq n\}, & \text{if } lub \text{ exists} \\ set\ t & \text{otherwise.} \end{cases}$*
  where $v = \{v_1, \cdots, v_n\}$.   ■

In the above *lub* refers to *least upper bound* in the ordered set $(\mathcal{T}_C^u, \preceq_{\mathcal{T}_C^u})$. In fact the definition is independent of γ, and we can consider $v \in [\![t]\!]$, instead of $v \in [\![t]\!]^\gamma$. However, the definition depends on inheritance.

Theorem 3 *For every type t and every $v : t$, $mst(v : t)$ exists and is unique, moreover $mst(v : t) \preceq_{\mathcal{T}_C} t$.*   ■

Object Database Instance

Roughly speaking, an *object database instance* over an object database schema $\mathcal{S}$ consists of giving an interpretation of $\mathcal{S}$, in a "semantic universe". Our semantic universe is the universe of partial functions. An important requirement in an object-oriented database is that the semantics has to be close to the meaning of the real world objects. Each real object has a state and a behavior. The state is the concrete object and the behavior is the set of allowed computations on the state. State and behavior are defined, respectively, by attributes and methods. We assume that classified object identities represent actually the real world stored objects. Thus, we have to interpret attributes and methods. This interpretation must take into account the above semantics of inheritance and subtyping. It must also support overriding. We interpret each derived attribute of a class c by a function defined on all objects classified in c. These functions are given by extension (i.e. by their graphs) and are supposed to be stored. They form a *state* of the database schema. Interpreting the behavioral schema is a more sophisticated problem. We have to associate each specified method with a calculation. To do this we need a language for coding methods. Coding methods can be realized in two ways:

- either by considering a traditional programming language like C, C++, or
- by designing a special language.

In the first case the programming language may have to be enriched with features supporting persistent and abstract types. Most objected-oriented database systems to date have chosen this route. We believe that, contrary to object programming languages, methods in object-oriented databases are simple calculations and operate only on concrete stored objects. Thus, we do not need a sophisticated programming language for coding methods. Moreover, methods and query languages must be able to manipulate null values (i.e. unknown values), and most classical languages lack this feature. Similarly to relational commercial query languages, simple calculations are also needed for querying. As a result a special and simple language $\mathcal{L}$ will have to be designed in Section 5. This language will serve to code methods and to assist the query language. Thus, it must support the type system presented in this paper. In fact $\mathcal{L}$ will be designed over a hierarchically consistent database schema. If $m : t_1 \cdots t_n \to t$ is a method in a class c, it will be implemented as a program in $\mathcal{L}$ with profile $c \otimes t_1 \otimes \cdots \otimes t_n \to t$ (recall that procedures are seen as functions with type *unit*). In what follows we consider a hierarchically consistent database schema $\mathcal{S} = (C, isa, att, meth)$ and such a language $\mathcal{L}$ over $\mathcal{S}$.

Definition 13 *An object-oriented database instance over the object-oriented database schema $\mathcal{S}$ and the language $\mathcal{L}$ is a tuple $\mathcal{I} = (\llbracket \rrbracket, \gamma, \alpha, \mu)$ such that*

- $\llbracket \rrbracket$ *is a type interpretation function and γ is a classification function over C,*
- α *is a function that associates each attribute (c, a, t) in $\overline{S}$ with a function $a_\gamma^c : \gamma(c) \to \llbracket t \rrbracket_\perp^\gamma$,*
- μ *is a function that associates each method $(m, t_1 \cdots t_n, t) \in meth(c)$ in $\mathcal{S}$ with a label $\iota \in \{u, s, d, ds\}$ and a program $m_{\iota, \gamma}^{c, r} : c \otimes r \to t$ in $\mathcal{L}$, where $r = t_1 \otimes \ldots \otimes t_n$. The label ι is u whenever m is a method with side effect.* ∎

The label ι indicates the kind of the method or its evaluation mode (u stands for *update*, s for *static*, d for *dynamic*, and ds for *dynamic-static*, see Section 5.3). Now, an *object database* over BASE, C, ATT and METH consists of a type system over BASE and C, a hierarchically consistent database schema $\mathcal{S}$ over that type system, and a database instance $\mathcal{I}$ over that schema. We say $(\llbracket \rrbracket, \gamma, \alpha)$ is the *state* and $(\llbracket \rrbracket, \gamma, \mu)$ is the *behaviour* of the database $\delta = (\mathcal{S}, \mathcal{I})$. The definition says that each derived attribute of a class must be implemented, but only explicitly specified methods of the class need to be implemented. The definition ensures that attributes cannot refer to objects outside the database instance. This is, in some way, the counterpart of a foreign key constraint in the relational language SQL. The superscripts indicate that the interpretation of an attribute depends on the class c in which that attribute is specified, while the interpretation of a method depends on c and on the type r of arguments of the specified method. The subscript γ in a_γ^c and $m_{\iota,\gamma}^{c,r}$ indicates that they depend on γ. The subscript ι says that the semantics of m depends on the evaluation mode (see Section 5.3). In fact, we will see, more precisely, that a_γ^c depends on c, γ and $[\![]\!]$; while $m_{\iota,\gamma}^{c,r}$ depends on c, r, γ, $[\![]\!]$ and α.

We refer to the above definition of object database instance as the *attributes-as-functions* approach. Let us give a comparison of this definition with the definition of relational database instance seen in Section 2, namely the "tuples-as-functions" approach. For any $o \in \gamma(c)$ let $v_i = a_{i_\gamma}^c(o)$ and define the function $v : \{a_1, \cdots, a_n\} \to \bigcup_{1 \leq i \leq n} [\![t_i]\!]_\perp^\gamma$ by $v(a_i) = v_i$ $(i = 1, \cdots, n)$. The function v is the counterpart of what we called a tuple in the relational model. The pair $\tilde{o} = (o, v)$ is called an *object* of c. The tuple v is the *value* of this object and o is its identity. Let us call a finite set of objects of c an *extension* of c. The set $\delta_\gamma^c = \{\tilde{o} \mid o \in \gamma(c)\}$ is an extension of c. If we denote the set of extensions of c by $\widetilde{\mathcal{E}xt}(c)$ then the state $([\![]\!], \gamma, \alpha)$ in Definition 13 defines a function $\tilde{\alpha} : C \to \bigcup_{c \in C} \widetilde{\mathcal{E}xt}(c)$ such that $\tilde{\alpha}(c) = \delta_\gamma^c$. This function $\tilde{\alpha}$ is the counterpart of the function α in Definition 2 of relational database instance.

We claim that, in a theoretical point of view, the "attributes-as-functions" approach presented in this section is more convenient for object oriented models, as it allows to deal with attribute and methods at the same level and from an algebraic view point (see Section 5).

For any class c we give a relational-like tabular representation of $(c, \overline{att}(c), \tilde{\alpha}(c))$ as shown in our concrete example in Appendix A. In this representation c is the name of the table, $\overline{att}(c)$ its heading row and $\tilde{\alpha}(c)$ the set of other rows.

4   An Algebraic Query Language for the Object Model

As we have seen earlier querying a database consists of extracting data from the database. These data represents the answer to a query on the real world objects. In the relational model, a query is specified and evaluated as an algebraic expression. Each query has a schema and the result of the evaluation of a query is an extension (a set of data) over its schema. For defining an algebraic query language for our object model we shall follow the same lines.

4.1   Query Specification

In the relational algebra each query q is associated with a set of attributes $att^*(q)$, called the query schema. The relational algebra, however, is unable to deal even with simple computations over data. In object models it is desirable that a query language manipulates computations over data. For instance, this is the case in the proposed standard language OQL [9]. In our framework, we see a computation as a functional program in the language $\mathcal{L}$ (we mean a program whose type is not *unit*). In the sequel such a program is called a *functional term*. At the syntactic level a functional term is seen as a function $f : t_1 \rightarrow t_2$ where $t_1 \in \mathcal{T}_C^u$ but $t_2 \in \mathcal{T}_C$. For example, side effect free methods are functional terms. In fact, as we shall see in Section 5, functional terms are expressions of an algebra. This algebra is generated by the type system and by other information in the database. In what concerns the query language, we assume that methods are side effect free. That is :

for each method $m : t_1 \cdots t_n \rightarrow t$ the type t is not *unit*.

Since overloading attributes and methods is allowed in an object schema and since computations can be used in queries we can not determine a schema for a query, as we did in the relational model. On the other hand, a query will be accompanied with a type called its *support*. The support of a query is the type of values belonging to its answer. The support of a query should not be confused with the type of the query. In this paper, we do not discuss the type of a query. Thus, support defines a function $supp : Q \rightarrow \mathcal{T}_C$, where Q is the set of all query expressions and $\mathcal{T}_C$ is the set of all type expressions. In the following we consider a hierarchically consistent database schema $\mathcal{S} = (C, isa, att, meth)$ and its closure $\overline{\mathcal{S}} = (C, isa, \overline{att}, \overline{meth})$.

Definition 14 *The set Q of queries and the function supp $: Q \rightarrow \mathcal{T}_C$ are defined recursively as follows:*

- *if $c \in C$ then $c \in Q$ and $supp(c) = c$,*
- *if $q \in Q$ and $f : supp(q) \rightarrow t$ is a functional term then $\mathtt{map}(f, q) \in Q$ and $supp(\mathtt{map}(f, q)) = t$,*
- *if $q \in Q$ and $f : supp(q) \rightarrow bool$ is a functional term (a condition over $supp(q)$) then $q[f] \in Q$ and $supp(q[f]) = supp(q)$,*
- *if $q \in Q$ and $f : supp(q) \rightarrow t_1$ and $g : supp(q) \rightarrow t_2$ are functional terms then $\mathtt{group}(f, g, q) \in Q$ and $supp(\mathtt{group}(f, g, q)) = t_1 \otimes (set\ t_2)$,*
- *if $q \in Q$ and $f : supp(q) \rightarrow set\ t$ is a functional term then $\mathtt{ungroup}(f, q) \in Q$ and $supp(\mathtt{ungroup}(f, q)) = supp(q) \otimes t$,*
- *if $q_1 \in Q$ and $q_2 \in Q$ then $q_1 \times q_2 \in Q$ and $supp(q_1 \times q_2) = supp(q_1) \otimes supp(q_2)$,*
- *if $q_1 \in Q$ and $q_2 \in Q$ and $supp(q_1) = supp(q_2) = t$ then $q_1 \cup q_2 \in Q$, $q_1 - q_2 \in Q$ and $supp(q_1 \cup q_2) = supp(q_1 - q_2) = t$,*
- *the only elements of Q are those obtained as above.* ∎

Note that the above definition uses the schema $\mathcal{S}$ and functional terms, only. The operations $\mathtt{map}(\text{-}, \text{-})$, $\text{-}[\text{-}]$, $\mathtt{group}(\text{-}, \text{-}, \text{-})$, $\mathtt{ungroup}(\text{-}, \text{-})$, $\text{-} \times \text{-}$, $\text{-} \cup$ - and - $-$ - are called $\mathtt{mapping}$, $\mathtt{selection}$, $\mathtt{grouping}$, $\mathtt{ungrouping}$, $\mathtt{product}$, $\mathtt{union}$ and $\mathtt{difference}$, respectively. They form a partial (universal) algebra that we call the *object algebra*.

Before giving the formal evaluation process of this query language let us see roughly its similarity with the relational algebra and its usual extensions. The operation map is a generalization of the projection operation as defined on the relational algebra, nested relational algebra and complex object algebra [29], [28], [2]. The projection is obtained when f is the projection function of a product type on some of its components. But in a more general setting f can contain path expressions using attributes or methods. The selection operation generalizes the selection operation as defined in the relational and nested relational models. Indeed, in $q[f]$ the function f can be a simple relational like condition or a more sophisticated condition using path expressions. The operation group gives a uniform formalization of the nesting operation ν of the nested relational model and the operation group by of SQL or OQL. The operation ungroup generalizes the unnesting operation μ of the nested relational model as well as those queries of OQL which perform an iteration in the clause from over a multi-value attribute (for instance in our example select ... from x in *Stud*, y in x.takes where ...). Operations product, union and difference generalize similar set-operations for the relational model and its extensions. Our object algebra has the closure property in the sense that query may itself be an argument of an operation (see Appendix A for concrete examples).

4.2   Query Evaluation

Evaluating a query q over a database instance $\mathcal{I} = (\llbracket \rrbracket, \gamma, \alpha, \mu)$ of $\mathcal{S}$ consists of extracting a particular finite subset of $\llbracket supp(q) \rrbracket^\gamma$ that we call the *answer* to q. For each type t let us denote the set $\mathcal{P}_f(\llbracket t \rrbracket^\gamma)$ by $\mathcal{E}xt(t)$. Then the evaluation process consists of defining a convenient function $\alpha^* : \mathcal{Q} \to \bigcup_{q \in \mathcal{Q}} \mathcal{E}xt(supp(q))$, such that $\alpha^*(q) \in \mathcal{E}xt(supp(q))$. This function α^* is the counterpart of the function α^* in Section 2 for the relational model. In order to define α^* we need to know how a functional term is evaluated. Because of subtyping and overloading the same functional term f (meaning the same algebraic expression) may be typed in different way. However, in the context of a query this kind of ambiguity can not happen. The rules for construction of functional terms and their semantics, as we will define them in Section 5, allow us to see each functional term $f : t_1 \to t_2$ at the semantic level as a function $\llbracket f \rrbracket_\delta^{\llbracket t_1 \rrbracket} : \llbracket t_1 \rrbracket^\gamma \to \llbracket t_2 \rrbracket_\perp^\gamma$. The notation indicates that this semantics depends on the semantics of type t_1 and on the database instance $\delta = (\mathcal{S}, \mathcal{I})$. In fact, the type t_2 "is inferred" from t_1 and f (see Section 5).

Definition 15 *The evaluation function $\alpha^* : \mathcal{Q} \to \bigcup_{q \in \mathcal{Q}} \mathcal{E}xt(supp(q))$ is defined recursively as follows :*

- $\forall \, c \in C, \; \alpha^*(c) = \llbracket c \rrbracket^\gamma = \bigcup_{c' \leq_{isa} c} \gamma(c')$

- $\alpha^*(\text{map}(f, q)) = \{ \llbracket f \rrbracket_\delta^{\llbracket t \rrbracket}(x) \mid (x \in \alpha^*(q)) \wedge (\llbracket f \rrbracket_\delta^{\llbracket t \rrbracket}(x) \neq \perp) \}$,

- $\alpha^*(q[f]) = \{ \llbracket f \rrbracket_\delta^{\llbracket t \rrbracket}(x) \mid (x \in \alpha^*(q)) \wedge \llbracket f \rrbracket_\delta^{\llbracket t \rrbracket}(x) \}$,

- $\alpha^*(\text{group}(f,g,q)) = \{(\llbracket f \rrbracket_\delta^{\llbracket t \rrbracket}(x), G) \mid x \in \alpha^*(q) \wedge (G = \{\llbracket g \rrbracket_\delta^{\llbracket t \rrbracket}(y) \mid$
$$y \in \alpha^*(q) \wedge (\llbracket g \rrbracket_\delta^{\llbracket t \rrbracket}(y) \neq \bot)\})\},$$
- $\alpha^*(\text{ungroup}(f,g,q)) = \{(x,y) \mid x \in \alpha^*(q) \wedge (\llbracket f \rrbracket_\delta^{\llbracket t \rrbracket}(x) \neq \bot) \wedge (y \in \llbracket f \rrbracket_\delta^{\llbracket t \rrbracket}(x))\},$
- $\alpha^*(q_1 \times q_2) = \alpha^*(q_1) \times \alpha^*(q_2)$,
- $\alpha^*(q_1 \cup q_2) = \alpha^*(q_1) \cup (q_2)$ and $\alpha^*(q_1 - q_2) = \alpha^*(q_1) \backslash \alpha^*(q_2)$,

where $t = supp(q)$ and f and g are typed as in Definition 14. ∎

Since the database is finite the evaluation of any query provides a finite set of tuples. Moreover, at least one of the components of each tuple is not the null value $\bot$. This is due to our semantics of $\otimes$ which is not the cartesian product. The first clause above is the basis of recursion. It ensures $\alpha^*(q) \in \mathcal{E}xt(supp(q))$, for each query q. Thus, α^* defines a function from $\mathcal{Q}$ to $\bigcup_{q \in \mathcal{Q}} \mathcal{E}xt(supp(q))$ (See Appendix A for concrete examples).

5   Functional Terms : Syntax and Semantics

Recall that a database schema $\mathcal{S} = (C, isa, att, meth)$ and its closure $\overline{\mathcal{S}} = (C, isa, \overline{att}, \overline{meth})$ are defined over a type system. This type system is built-up over basic types and a finite set C of class names. Basic types are usually accompanied with predefined operations. In this Section we will construct the syntax of our language $\mathcal{L}$ over a hierarchically consistent schema $\mathcal{S}$, and the semantics of $\mathcal{L}$ with respect to a database δ over $\mathcal{S}$. Terms of this language are called *functional terms*. At the semantic level each functional term e is interpreted as a partial function. Therefore, we display e at the syntactic level as $e : t \to t'$. The type t ranges over $\mathcal{T}_C^u$ while t' ranges over $\mathcal{T}_C$.

5.1   Syntax of Functional Terms

The following rules construct functional terms from some constant terms, attributes and methods declared in the schema $\mathcal{S}$.

(category)	$\dfrac{}{id : t \to t}$	$\dfrac{e : t \to t' \quad f : t' \to t''}{e.f : t \to t''}$	
(product)	$\dfrac{}{fst : t_1 \otimes t_2 \to t_1}$	$\dfrac{}{snd : t_1 \otimes t_2 \to t_2}$	$\dfrac{e_1 : t \to t_1 \quad e_2 : t \to t_2}{< e_1, e_2 >\, : t \to t_1 \otimes t_2}$
(partiality)	$\dfrac{f : t \to t'}{isnul(f) : t \to bool}$	$undef_t : unit \to t$	$ter : t \to unit$
(basic types)	$\dfrac{v : b}{v : unit \to b}$	$\dfrac{op : b_1 \cdots b_n \to b}{op : b_1 \otimes \cdots \otimes b_n \to b}$	$b, b_1, \cdots, b_n \in \text{BASE}$
(conditional)	$\dfrac{}{if - then - else - \,: bool \otimes t \otimes t \to t}$		
(database)	$\dfrac{\overline{att}(c,a) = t}{a^c : c \to t}$	$\dfrac{t \preceq t'}{sub^{t,t'} : t \to t'}$	$c \in C,\ a \in \text{ATT}$
	$\dfrac{meth(c,m,r) = t}{m_\iota^{c,r} : c \otimes r \to t}$	$\iota \in \{s, d, ds\},\ c \in \text{C},\ m \in \text{METH},\ r \in \mathcal{T}_C^*$	

In the rules above, b is a basic type and $\underline{op}$ is a usual operation on basic types (which supposed to be predefined). Functional terms are expressions obtained from the above rules. The same functional term e may have different sources or different targets, because of overloading methods and attributes. However, the following fact implies that there is no confusion.

Fact 6 *There is only one functional term with the same expression and the same source, that is if $e_1 : t_1 \to t_2$ and $e'_1 : t'_1 \to t'_2$ are two functional terms such that $e_1 = e'_1$ (syntactically) and $t_1 = t'_1$ (syntactically) then $t'_2 = t_2$ (syntactically).*

Proof : By structural recursion. ∎

5.2   Semantics of Functional Terms

Let $\mathcal{S}et$ be the usual category of sets and (total) functions with its usual cartesian product "$\times$" and its usual co-product "$+$" (disjoint union). We denote by $< \_, \_ >$ and $[\_, \_]$ the pairing and co-paring combinators in this category, respectively. Therefore, for functions $f : a \to b$ and $g : a \to c$ the function $< f, g >: a \to b \times c$ is defined by $< f, g > (x) = (f(x), g(x))$. Similarly, for functions $h : a \to c$ and $k : b \to c$ the function $[h, k]$ is defined by $[h, k](x) = h(x)$ when $x \in a$ and by $[h, k](x) = k(x)$ when $x \in b$. Projections of $a \times b$ are denoted $pr_1^{a,b}$ and $pr_2^{a,b}$, and co-projections of $a + b$ are denoted $in_1^{a,b}$ and $in_2^{a,b}$. If $f : a \to b$ and $g : c \to d$ then $< f \circ pr_1^{a,c}, g \circ pr_2^{a,c} >$ is denoted $f \times g$ and $[in_1^{b,d} \circ f, in_2^{b,d} \circ g]$ is denoted $f + g$. The above notations can be generalized for a product or sum of n factors, mutatis-mutandis

For any set a, we denote the set $a + \{\bot\}$ by $a_\bot$ and the co-projections of $a + \{\bot\}$ by $left^a : a \to a_\bot$ and $right^a : \{\bot\} \to a_\bot$. Each $u : a \to b_\bot$ defines the function $ext\ u = [u, right^b] : a_\bot \to b_\bot$. In fact, the function $ext\ u$ extends u to $\bot$ by $(ext\ u)(\bot) = \bot$. If x is an element of a set a we have a function $\tilde{v} : \{\bot\} \to a$ defined by $\tilde{v}(\bot) = v$. For instance, $\widetilde{\mathbf{true}} : \{\bot\} \to bool$ and $\widetilde{\mathbf{false}} : \{\bot\} \to bool$ represent truth values functionally. There is also a unique function $\widetilde{ter}^a : a \to \{\bot\}$. If $a \subseteq b$ then the inclusion function is denoted $incl^{a,b} : a \hookrightarrow b$.

Given an object database δ we will define a function $[\![\ ]\!]_\delta$ which interprets each functional term as a partial function. But, in this section a partial function f^* from a to b will be seen as a usual function $f : a \to b_\bot$ where $f(x) = \bot$ means f^* is not defined in x. Now, let $\mathcal{I} = ([\![\ ]\!], \gamma, \alpha, \mu)$ be a database instance over $\mathcal{S}$. The semantics of functional terms with respect to a database instance $\delta = (\mathcal{S}, \mathcal{I})$ is defined as in the boxes below.

$$[\![if - then - else-]\!]_\delta^{t_2} = [[\![fst]\!]^{t_2} \circ snd^{bool,[\![t2]\!]^\gamma}, [\![snd]\!]^{t_2} \circ snd^{bool,[\![t2]\!]^\gamma}]$$
$$+ [\![undef_t]\!] \circ \widetilde{ter}^{[\![unit]\!],[\![t2]\!]^\gamma}$$
$$+ \widetilde{ter}^{[\![unit]\!],[\![unit]\!] \times [\![unit]\!]} \quad : bool \otimes t \otimes t \to t$$
$$\text{where } t_1 = [\![bool \otimes t \otimes t]\!], \ t_2 = [\![t \otimes t]\!]$$

$$\llbracket id \rrbracket_\delta^{[t]} = left^{[t]^\gamma} : [t]^\gamma \to [t]_\perp^\gamma$$

$$\llbracket e.f \rrbracket_\delta^{[t]} = ext(\llbracket f \rrbracket_\delta^{[t]}) \circ \llbracket e \rrbracket_\delta^{[t]} : [t]^\gamma \to [t'']_\perp^\gamma$$

$$\llbracket fst \rrbracket_\delta^{[t_1 \otimes t_2]} = [pr_1^{[t_1]^\gamma,[t_2]^\gamma}, pr_1^{[t_1]^\gamma,[unit]^\gamma}] + pr_1^{[unit]^\gamma,[t_2]^\gamma} : [t_1 \otimes t_2]^\gamma \to [t_1]_\perp^\gamma$$

$$\llbracket snd \rrbracket_\delta^{[t_1 \otimes t_2]} = [pr_2^{[t_1]^\gamma,[t_2]^\gamma}, pr_2^{[unit]^\gamma,[t_2]^\gamma}] + pr_2^{[t_1]^\gamma,[unit]^\gamma} : [t_1 \otimes t_2]^\gamma \to [t_2]_\perp^\gamma$$

$$\llbracket < e_1, e_2 > \rrbracket_\delta^{[t]} = < \llbracket e_1 \rrbracket_\delta^{[t]}, \llbracket e_2 \rrbracket_\delta^{[t]} >: [t]^\gamma \to [t_1 \otimes t_2]_\perp^\gamma$$

$$\llbracket isnull(f) \rrbracket_\delta^{[t]} = (left^{[bool]} \circ [\widetilde{true} \circ \widetilde{ter}^{[t]^\gamma}, false] \circ \llbracket f \rrbracket_\delta^{[t]}) : [t]^\gamma \to [bool]_\perp^\gamma$$

$$\llbracket undef_t \rrbracket_\delta^{[unit]} = right^{[t]^\gamma} : [unit]^\gamma \to [t]_\perp^\gamma$$

$$\llbracket ter \rrbracket_\delta^{[t]} = (left^{[unit]^\gamma} \circ \widetilde{ter}^{[t]^\gamma}) : [t]^\gamma \to [unit]_\perp^\gamma$$

$$\llbracket v \rrbracket_\delta^{[c]} = \tilde{v} : [unit]^\gamma \to [b]_\perp^\gamma,$$

$$\llbracket sub^{t,t'} \rrbracket_\delta^{[t]} = incl^{[t]^\gamma,[t']_\perp^\gamma} : [t]^\gamma \hookrightarrow [t']_\perp^\gamma$$

$$\llbracket op \rrbracket_\delta^{[b_1 \otimes \cdots \otimes b_n]} = < left^{[b]} \circ \llbracket op \rrbracket, \nu_2, \cdots, \nu_k >: [b_1 \otimes \cdots \otimes b_n]^\gamma \to [b]_\perp^\gamma$$

$$\llbracket a^c \rrbracket_\delta^{[c]}(x) = \begin{cases} (ext(\llbracket sub^{t',t} \rrbracket[t']) \circ a_\delta^{c'})(x) & \text{if } x \in \gamma(c'), c' \leq_{isa} c, \\ & \text{and } \overline{att}(c',a) = t' \\ \perp & \text{otherwise} \end{cases}$$

Some clauses of this semantics need to be clarified.

The definition $\llbracket if - then - else - \rrbracket^{bool \otimes t \otimes t}$ says that:

 - if the first argument is **true**, then the result is the second argument,
 - if the first argument is **false**, then the result is the third argument,
 - if the first argument is $\perp$, then the result is $\perp$ too.

In $\llbracket op \rrbracket_\delta^{[b_1 \otimes \cdots b_n]}$, by definition of the semantics of $\otimes$ the set $[b_1 \otimes \cdots \otimes b_n]$ is a sum of $k = 2^n - 1$ factors. Each of these factors is a usual product of n sets from $[b_1], \cdots, [b_n], [unit]$ excluding $[unit]^n$. The first factor is $[b_1] \times \cdots \times [b_n]$. We denote $\nu_i = right^{[b]} \circ \widetilde{ter}^{N_i}$ where N_i is one of the other factors. The usual semantics of the basic operation op is seen as a function $\llbracket op \rrbracket : [b_1] \times \cdots \times [b_n] \to [b]$. More precisely, $\llbracket op \rrbracket^{b_1 \otimes \cdots \otimes b_n}$ is defined as $\llbracket op \rrbracket$ whenever non of its arguments is $\perp$ and is undefined otherwise.

In $\llbracket a^c \rrbracket_\delta^c$, if $c' \leq_{isa} c$ and $\overline{att}(c',a) = t'$ then $t' \preceq t$ (by covariance), $a_\gamma^{c'} : \gamma(c') \to [t']_\perp^\gamma$ (by definition of δ), and $\llbracket sub^{t',t} \rrbracket[t'] : [t']^\gamma \to [t]_\perp^\gamma$. Thus, $\llbracket a^c \rrbracket_\delta^{[c]}$ is a well-defined function. It is not difficult (but tedious) to write $\llbracket a^c \rrbracket_\delta^{[c]}$ with categorical combinators.

To complete the semantics we have to define $\llbracket m_{iota}^{c,r} \rrbracket$ which needs more detailed explanations. Indeed, the semantics of a method depends on the chosen evaluation mode as explained in the next subsection.

5.3   Evaluation Modes for Methods

In a schema, a method name m can appear in many classes or several times in the same class. In the database instance each occurrence of m is associated with a functional term. Invoking a method m on an argument w needs binding one of these functional terms to w. The functional term which must be bound depends on the evaluation mode we choose. As mentioned earlier, the database instance associates each method $m : r \to t$, in a class c, with a functional term $m_{i,\gamma}^{c,r} : c \otimes r \to t$ where $r = t_1 \otimes \cdots \otimes t_n$. We write a value of $c \otimes r$ as (o, v) where $v = (v_1, \cdots, v_n)$. The problem is how to evaluate m on (o, v), namely

$o.m(v)$? We will see two kinds of binding which include the binding modes of object programming languages especially C++ and Java. Indicating the binding mode we will denote the name m of a method by m_ι where $\iota \in \{s, d, ds\}$ (s for static, d for dynamic, ds for dynamic-static).

Early Binding or Static Evaluation

The type of an object in this mode is its static type (declared type), which is known at compile time. Let $(o, v) : c \otimes r$ and $(o, v) \in [\![c \otimes r]\!]^\gamma$. For an explicit method $m : r \to t$ in c we define $o.m_s(v) = [\![m_{s,\gamma}^{c,r}]\!]_\delta^{[\![c \otimes r]\!]}(o, v)$. In particular, if o is not γ-identifiable in c or if v is not γ-identifiable in r then $o.m_s(v) = \bot$. If m is a derived method in c then by hierarchical consistency $ResM(c, m, r) = (c', r')$, $c \otimes r \preceq_{T_C} c' \otimes r'$ and $m : r' \to t$ is an explicit method in c'. In this case we define $o.m_s(v) = ext(m_\delta^{c',r'}) \circ [\![sub^{c \otimes r, c' \otimes r'}]\!]_\delta^{c \otimes r}$. It is clear that $(o, v) \mapsto o.m_s(v)$ defines a function from $[\![c \otimes r]\!]^\gamma$ to $[\![t]\!]_\bot$ that we denote by $[\![m_s^{c,r}]\!]_\delta^{[\![c \otimes r]\!]}$. This function is the semantics of $m_s^{c,r}$ for the early binding mode.

Late Binding or Dynamic Evaluation

The type of an object in this mode is its dynamic type, that we called the most specific type (Section 3.4). In object programming languages there are two main dynamic binding modes, namely multiple dispatching and simple dispatching.

Multiple Dispatching : Consider an explicit or derived method $m : r \to t$ in c and a value $(o, v) : c \otimes r$. Define the value $o.m_d(v)$ by the following algorithm:

```
if o is not γ-identifiable in c or v is not γ-identifiable in r
then
    o.m_d(v) = ⊥
else
    (c_1, r_1) = mst((o, v) : c ⊗ r)
    (c_2, r_2) = ResM(c_1, m, r_1)
    o.m_d(v) = ⟦m_{d,γ}^{c_2,r_2}⟧^{⟦c_2⊗r_2⟧}(o, v).
```

Fact 7 *The assignment $(o, v) \mapsto o.m_d(v)$ defined by the above algorithm gives a well-defined function $[\![m_d^{c,r}]\!]_\delta^{[\![c \otimes r]\!]} : [\![c \otimes r]\!]^\gamma \to [\![t]\!]_\bot^\gamma$. This function is the semantics of $m_d^{c,r}$ for the late binding mode with multiple dispatching.* ∎

This fact says that multiple dispatching is type-safe, meaning, it doesn't cause type errors in run time. However, rare are the object programming languages that use this mode. The reason is the high cost of resolution of overloading problems. We believe that in object database systems this mode is the most appropriate. Indeed, in object databases the schema evolves less quickly than in programming languages. Thus, we can store the whole closure of a schema which contains all information for binding. Moreover, run time errors in a database context fires the heavy artillery of crash recovery and database restoration.

Single Dispatching : In this mode evaluating $o.m_{ds}(v)$ needs the dynamic type of o but the static type of v. In our approach the value $o.m_{ds}(v)$ is defined by the following algorithm :

> if o is not γ-identifiable in c or v is not γ-identifiable in r
> then
> $\quad o.m_{ds}(v) = \bot$
> else
> $\quad (c_1, r_1) = ResM(c, m, r)$
> $\quad c_2 = mst(o : c)$
> $\quad (c_3, r_3) = ResM(c_2, m, r_1)$
> $\quad o.m_{ds}(v) = [\![m_{ds}^{c_3, r_3}]\!]^{[\![c_3 \otimes r_3]\!]}(o, v).$

Fact 8 *The assignment $(o, v) \mapsto o.m_{ds}(v)$ defined by the above algorithm gives a well-defined function $[\![m_{ds}^{c,r}]\!]_{\delta}^{[\![c \otimes r]\!]} : [\![c \otimes r]\!]^{\gamma} \to [\![t]\!]_{\bot}^{\gamma}$. This function is the semantics of $m_{ds}^{c,r}$ for the late binding mode with single dispatching .* ∎

Other kinds of late binding may be considered. Instead of taking static types for v we can imagine to consider dynamic types for some components of v and static types for others.

Binding Modes in other Models : Alternative strategies for early or late binding exist in most object languages and systems.

– In O2 [20] early binding depends only on the static type of the invoking object. Late binding in O2 depends only on the dynamic type of the invoking object. Ignoring dynamic and static types of parameters seems to be the source of some run time errors in O2.
– In Java the default mode is late binding. The keyword FINAL enforces the static mode. The early binding mode of JAVA seems be the same as ours, but late binding is the single dispatching mode. That is, it depends on the dynamic type of the invoking object and on the static types of the parameters (see [27] for examples).
– In C++ the default mode is static binding. The keyword VIRTUAL enforces dynamic binding. Like JAVA, C++ ignores dynamic types of parameters.

6   Related Works

A data model with a clear separation between schema and state is introduced in [22], [24] and carried on in [23] and [30]. The main characteristic of these works is their algebraic aspect. The schema is defined similar to an algebraic specification of data types and the database state as an algebra of that specification. An algebraic query language for this model has been sketched in [23] and further developed in [30]. The present work completes and enrich the latter and gives more formalization.

Our work is influenced by two other approaches : the approach adopted in the database system O2 [20], [4] and work in connection with collection types [7], [5],

[6], [13], [12]. The first approach resulted in the proposal of ODL model and the language OQL by the ODMG group [9], [10], [11]. It allowed to point out the basic operations on objects and to better understand the border between object databases and object programming languages. The second approach allowed to see the similarity of techniques from databases and from functional programming. The main idea of the approach presented in [7], [5], [6] is to enrich the functional programming language techniques with facilities which allow to manipulate collections. The approach proposes an algebra and a calculus. The algebra manipulates total functions only. As a result the approach cannot deal with null values. In contrast to this approach, ours uses partial functions and includes null values, which is quite typical in practical applications. Some of our rules for defining the language $\mathcal{L}$ in Section 5.1 are similar to their rules. However, our semantics of these rules are completely different.

The same idea of collection types in terms of an algebraic structure, called monoid, is introduced in [13], [12]. In this approach a type is a monoid and a query is a monoid homomorphism. As a consequence this approach captures multiple collections in the same database. In the query language a value null appears, however it is not clear what its semantics is and how it can appear at the schema level. In contrast, the approach of [13], [12] enables to manipulate aggregate functions as do SQL and OQL, while our approach can not. However, our model can be enriched with such manipulations, as shown in [30].

Both of the above approaches propose a calculus, and use a strong degree of integration of database schema, database instance and query language. They don't address the issues of inheritance, overloading and methods. In contrast, our approach lacks a calculus but includes all aspects of objects, and has three distinct levels : schema, instance and query. Moreover, we provide a nice formalization of binding modes in object databases and object programming languages.

7   Conclusions and Future Works

We have introduced a formal object-oriented data model and its semantics in the style of the relational model. We have shown that our model supports all important aspects of objects namely inheritance, overriding, and early and late binding. We have deduced a subtyping relation from inheritance and we have used it for giving a nice formalization of delicate problems such as resolution, multiple inheritance, covariance and binding modes. We claim that this formalization allows to better understand object models and especially to clarify the difference between object databases and object programming languages. Attributes and methods have been presented at the semantic level as partial functions. But partiality is treated in this paper as undefinedness rather than as program failure, as it is usually considered in computation theory. We have considered a partial function $f : t \rightarrow t'$ as a total function $t \rightarrow t' \cup \{\bot\}$ where $\bot$ is outside t' and represents the null value of databases. The underlying type system of our model is endowed with a particular constructor $\otimes$. The semantics of $\otimes$ have been defined as $t \otimes t' = t \times t' + \{\bot\} \times t' + t \times \{\bot\}$ and not as the usual cartesian product. We proved that this semantics suits with partiality and is convenient for dealing with objects in database concepts in which only few, but not all attribute values of an object may be unknown.

Several particularly interesting aspects of the model have not been presented here including query optimization, methods with side effects, and multiple collections. The first aspect is sketched in [23] and [30] while the two other aspects need deeper investigation. Further work should also consider the connection of this model with category theory and monads. In fact the partial functions as treated in this paper should be seen as arrows in the Kleisli category of a monad (the lifting monad) and $\otimes$ as a product in that category. It would be interesting to study the connection of our query language with other algebraic query languages for object models in particular with the monoid calculus [13], [12]. Results in these directions will be reported in a forthcoming paper.

Acknowledgment. I am very grateful to George Loizou for his comments on a previous version of this paper. Many thanks to Nicolas Spyratos for his help and his precious comments on the final version. Thanks to Rachid Souah and Denis Bechet for their help especially with respect to binding modes.

References

[1] M. Abadi and L. Cardelli. *A Theory of Objects. Springer-Verlag*, 1996.

[2] S. Abiteboul and C.Berri. On the Power of languages for Manipulation of Complex Objects. *In Proceedings of International Workshop on Theory and Applications of Nested Relations and Complex Objects*, 1988.

[3] S. Abiteboul and R. Hull and V. Vianu. *Foundation of Databases, Addison-Wesley*, 1995.

[4] F. Bancilhon and S. Cluet and C. Delobel. A query language for O2. *In François Bancilhon, Claude Delobel, and Paris Kanellakis, editors, Building an Object-Oriented Database System, The Story of O2. Morgan Kaufmann*, 1992.

[5] V. Breazu-Tannen and P. Buneman and S. Naqvi. Structural Recursion as a Query Language. *In Proceedings of the Third International Workshop on Database Programming Language: Bulk types and Persistent Data*, pages 9-19, 1991.

[6] V. Breazu-Tannen and R. Subrahmaniam. Logical and Coputational Aspects of Programming with Sets/Bags/Lists. *In 18th International Colloquium on Automata, Languages and Programming, Madrid, Spain, Springer Verlag, LNCS 510*, pages 60-75,1991.

[7] P. Buneman and S. Naqvi and V. Tannen and Limsoon Wong. Principle of Programming with Complex Objects and Collection types. *Theorical computer Science*, 149:3-48, 1995.

[8] G. Castagna. *Object-Oriented Programming. A Unified Foundation, Birkhäuser*, 1997.

[9] R. Cattel. *The Object Databases Standard: ODMG-93, Release 1.2. Morgan Kaufmann*, 1996.

[10] S. Cluet. Designing OQL : Allowing object to be queried. *Information Systems* 23(5), pages 279-305, 1998.

[11] S. Cluet and G. Moerkeotte. Nested Queries in Object Bases. *In Fifth International Workshop on database Programming Languages*, pages 226-242, 1993.

[12] L. Fegaras. Query Unnesting in Object-Oriented Databases. *Proceedings of the ACM SIGMOD International Conference on Managmnenet of data, Seattle, Washington*, pages 49-60, 1998.

[13] L. Fegaras and D. Maier. Towards an Effective Calculus for Object query languages. *In Proc. of ACM SIGMOD International Conference on Management of Data*, pages 47-58,1995.

[14] J. Frohn and G. Lausen and H. Uphoff. Access to Objects by Path Expressions and Rules. *Proceedings of the 20th VLDB Conference, Santiago*, pages 273-284, 1994.

[15] R. Herzig and M. Gogolla. A SQL-like Query Calculus for Object-Oriented Database Systems. *International Symposium Object-Oriented Methodologies and Systems, LNCS 858*, pages 20-39, 1994.

[16] M. Kifer and W. Kim and Y. Sagiv. Querying Object-oriented databases. *In Proc. of the ACM SIGMOD Conference on Management of Data*, pages 393-402, 1992.

[17] W. Kim. *Modern Database Systems. The Object Model, Interoperability, and Beyond, Addison-Wesley*, 1995.

[18] W. Kim. A Model of queries for Object-orienrted databases. *In Proc. of the Intl. Conference on VLDB*, pages 423-432, 1989.

[19] G. Lausen and G. Vossen. *Models and Languages of Object-Oriented Databases, Addison-Wesley*, 1997.

[20] C. Lecluse and P. Richard and V. Velez. The O2 Data Model, *In François Bancilhon, Claude Delobel, and Paris Kanellakis, editors, Building an Object-Oriented Database System, The Story of O2 Morgan Kaufmann*, 1992.

[21] S.K. Lellahi. Type de collection et Monades. *Acte des Journées Catégories, Algèbres, Esquisses et Neo-esquisses, CAEN*, pages 109-114, 1994.

[22] S.K. Lellahi and N. Spyratos. Towards a Categorical data model Supporting Structural objects and Inheritance. *East/West Database Workshop, LNCS 504*, 1991, pp 86-105.

[23] S.K. Lellahi and R. Souah and N. Spyratos. An algebraic query language for Object-Oriented data Models. *DEXA97, LNCS N0 1308*, pages 519-528, 1997.

[24] S.K. Lellahi and N. Spyratos and M'B. Thiam. Functional Dependencies and the Semantics of Class extensions. *Journal of Computing and Information, Special issue: Proceedings of 8th ICCT* 2(1), pages 892-913, 1996.

[25] S.K. Lellahi and V. Tannen. A calculus for collections and aggregates. *Category and Computer Science, LNCS 1290*, pages 261-280, 1997.

[26] M. Levene and G. Loizou. *A Guided Tour of relational Databases and Beyond, Springer*, 1999.

[27] P. Niemeyer and J. Peck. *Exploring Java. O'Reilly & Associates Inc*, 1996.

[28] M.A. Roth and H.F. Korth and A. Silberschatz. The theory of Non-First-Normal-form Relational databases. TR-84-36, University of Texas at Austin, 1986.

[29] H.J. Schek and M.H. Sholl. The Relational Model with Relational valued Attributes. *Information Systems*, 11(2):137-147.

[30] R. Souah. *Une Sémantique Algébrique pour Bases de données Orientées objet.* PhD thesis, Université Paris-sud (Orsay), 1999.

[31] M'B. Thiam. *Dépendances Fonctionnelles et Consistance pour Base de données objet, PhD thesis, Université Paris 1 Panthéon-Sorbonne*, 1998.

[32] Y. Vassiliou. Null Values in Database Management : a Denotational Semantics Approach. *Proceedings of ACM SIGMOD International Conference on Management of Data*, pages 162-169, 1979.

[33] C. Zaniolo. Database Relations with Null Values. *Journal of Computer and System Sciences*, 28:142-166, 1984.

Appendix

8   An Object-Oriented Database

8.1   The Database Schema

```
class Pers
   attributes
      name : char
      spouse : Pers
      adr : Adr;

class Emp inherit Pers
   attributes
      salary : int
      adr : J_Adr
   methods
      bonus : → int;

class Dir inherit Emp
   attributes
      app_date : date
   methods
      seniority : date → int;

class Rworker inherit Emp
   attributes:
      lab : char
      prom_date : date
      adr : JE_Adr
   methods
      seniority : date → int;

class Prof inherit Emp
   attributes
      teaches : set  Course
      position : char
      e_adr : E_Adr
   methods
      bonus : int → int;

class Stud inherit Pers
   attributes
      super : Prof
      takes : set  Course;

class Tutor inherit Stud , Emp
   methods
      bonus : → int;
```

```
class Course
   attributes
      name : char
      level :int
      preq : set  Course;

class Adr
   attributes
      nbr : int
      street : char
      zip : Zip
   methods
      eq : Adr → bool;

class Zip
   attributes:
      town : char
      code : char⊗ int
      country : char
   methods
      eq : Zip → bool
      eq : char  char → bool;

class J_Adr inherit Adr
   attributes
      tel : int
      fax : int
   methods
      eq:J_Adr  → bool;

class E_Adr inherit Adr
   attributes
      e_mail : char
      web : char;
   methods
      eq : E_Adr → bool;

class JE_Adr inherit E_Adr, J_Adr
methods
      eq : JE_Adr → bool;
```

Comments : Basic types are `int`, `char`, `date` and `bool`. The attribute `name` is overloaded, the attribute `adr` of `Emp` is overridden in `Rworker`. The method `eq` is overloaded twice in `Zip` and the method `eq` of `Adr` is overridden in `J_Adr`, `E_Adr`, and `JE_Adr`. This schema is hierarchically consistent. For example

$$ResA(\mathtt{Prof}, \mathtt{adr}) = \mathtt{Emp}, \ ResM(\mathtt{E\_Adr}, \mathtt{eq}, \mathtt{J\_Adr}) = (\mathtt{Adr}, \mathtt{Adr}),$$
$$\overline{att}(\mathtt{Prof}, \mathtt{adr}) = \mathtt{J\_Adr}, \ \overline{meth}(\mathtt{E\_Adr}, \mathtt{eq}, \mathtt{J\_Adr}) = \mathtt{bool}.$$

8.2 The Database State

Pers	name	spouse	adr

Stud	name	spouse	adr	super	takes
8	Maria	1	14	4	{12, 13}
9	Sara	3	1	5	{12}

Emp	name	spouse	adr	salary
1	Remy	8	23	13000
2	Katy	10	24	14000

Dir	name	spouse	adr	salary	app_date
10	jean	2	24	15000	01/01/99

Rworker	name	salary	lab	adr	prom_date	spouse
3	Daniel	20000	e.n.s.	24	01/09/98	9

Prof	name	spouse	adr	e_adr	salary	position	teaches
4	Paul	⊥	23	22	13000	FullProf	{11, 12}
5	Yve	6	20	⊥	15000	AssProf	{13}
6	Eve	5	20	21	⊥	FullProf	{}

Tutor	name	spouse	adr	salary	super	takes
7	joseph	⊥	24	⊥	5	{11, 12}

Course	name	level	preq
11	bd	203	{12, 13}
12	system	101	{}
13	math	102	{}

Adr	nbr	street	zip
14	1	⊥	17
15	2	Ecole	18
16	13	St Michel	19

Zip	town	code	country
17	Piza	⊥	Italy
18	Paris	PA75006	France
19	Paris	PA75005	France

E_Adr	e_mail	web	nbr	street	zip
20	dl@up1.fr	http::://www....	2	Ecole	18
21	kl@up7.fr	http::://www....	20	Monge	19
22	⊥	http::://www....	⊥	⊥	17

J_Adr	nbr	street	zip	tel	fax
23	4	⊥	17	⊥	⊥
24	⊥	Ecole	18	0145920214	0145859246

JE_Adr	nbr	street	zip	tel	fax	e_mail	web
25	10	Roma	17	⊥	⊥	mt@coira.com	⊥
26	30	Zola	19	0145965402	0145647354	jd @ up1.fr	http::://www...

Comments : For simplicity Oid is defined as natural numbers.

$\gamma(\text{Pers}) = \emptyset,\ \gamma(\text{Emp}) = \{1,2\},\ \gamma(\text{Stud}) = \{8,9\}\ ,\ \cdots$

$[\![\text{Stud}]\!]^\gamma = \{7,8,9\},\ [\![\text{Pers}]\!]^\gamma = \{1,2,3,4,5,6,7,8,9,10\},\ \cdots$

$\alpha(\text{Stud},\text{name},\text{char}) = \text{name}^{\text{stud}}_\gamma : \{8,9\} \to [\![\text{char}]\!]_\perp$ is the function

$8 \mapsto \text{Marie}, 9 \mapsto \text{Sara}$.

8.3   The Database Behaviour

The following table implement methods by functional terms.

Class c	Method $m : r \to t$	Functional term $m^{c,r}_{\iota,\gamma} : r \to t$
Emp	bonus : → int	< ter. 0.05, salary >.mul_int
Dir	seniority : date → int	< snd, fst.app_date>.sub_date
Rworker	seniority : date → int	< snd, fst.prom_date>.sub_int
Prof	bonus : int → int	<< fst.ter.0.05, fst.salary>.mul_int, snd>.add_int
Tutor	bonus : → int	5000
Adr	eq : Adr → bool	<fst. nbr, snd.nbr>.eq_int and <fst. street, snd.street>.eq_char and <fst. zip, snd.zip>. eq_Zip
Zip	eq : Zip → bool	<fst. town, snd.town>.eq_char and <fst. code, snd.code>.eq_char⊗int and <fst. country, snd.country>.eq_char
Zip	eq : char char → bool	<fst. town, snd.fst>.eq_char and <fst. country, snd.snd>.eq_char
J_Adr	eq : J_Adr → bool	<fst. fax, snd. fax>.eq_int
E_Adr	eq : E_Adr → bool	<fst. e_mail, snd. e_mail>.eq_char and <fst. web, snd. web>.eq_char
JE_Adr	eq : JE_Adr → bool	<fst. fax, snd.fax>.eq_int and <fst. web, snd. web>.eq_char

Comments : In the above arithmetic operations $+, *, \cdots$ on int are denoted with prefix notations as add_int, mul_int, $\cdots$. Similarly equality on a type t is denoted eq_t. Methos have the following meaning:

The method bonus in Emp calculates 5 percent of salary, while the method bonus in Prof adds 5 percent of the salary to the argument. The method bonus in Tutor is a constant value. The method `seniority` in `Dir` gives the seniority of a director from his appointment date, while `seniority` in `Rworker` gives the seniority of a research worker since his last promotion date. The method `eq` in `Adr` checks whether two addresses are equal. The method `eq` in `J_Adr` checks whether two job addresses have the same fax. The method `eq` in `JE_Adr` checks whether two addresses have the same fax and the same web site. The first method `eq` in `Zip` checks whether zips are equal, whereas the second `eq` in `Zip` checks whether two zips concern the same town and the same country.

8.4   Some Examples of Queries

The following are some query expressions and their answers on our database state. The evaluation mode is the static mode.

q1	map((<name, salary>, Emp)	
Supp(q1)	char⊗ int	
$\alpha^*(q1)$	Remy	13000
	Katy	14000
	Daniel	20000
	Paul	13000
	Yve	15000
	Eve	⊥
	Joseph	⊥
	Jean	15000

q2	map((adr.zip.country, Pers)
Supp(q1)	char
$\alpha^*(q2)$	France
	Italy

q3	map((<name, spouse.name, position>, Prof)	
Supp(q3)	char⊗ char ⊗ char	
$\alpha^*(q3)$	Paul ⊥	Full Prof
	Yve Eve	Ass. Prof
	Eve Yve	Full Prof

q4	map((<name, e_adr.e-mail>, Prof[f])
Supp(q4)	char⊗ int
$\alpha^*(q4)$	Paul ⊥

where f is the functional term `<adr.zip.town, ter.Piza>.eq_char`

q5	group((adr.zip.country, name, Pers)
Supp(q5)	char⊗ set char
$\alpha^*(q5)$	France { Katy, Daniel, Yve, Eve, Joseph, Jean}
	Italy { Remy, Paul, Maria}

q6	group(fst, snd, q6)
Supp(q6)	char⊗ set int
$\alpha^*(q6)$	Bd {101, 102}

where
q6 = map((<name, snd.snd.level>,
ungroup(preq, Course)).

Comment : These queries have the following meaning on our database
q1 : Find the name and salary of all employees.
q2 : Find the country of origin of all persons.
q3 : Find the name, spouse name and position of each professor.

q4 : Find the name and e-mail of each professor who lives in Piza.

q5 : Find the name and country of origin of all persons, grouped with respect to their country.

q6 : Find for each course, levels of its prerequisites ?

8.5   Examples of Evaluation Modes

object	value	$o.m_s(v)$	$o.m_{ds}(v)$	$o.m_d(v)$
20	21	$eq_\delta^{Adr,Adr}(20,\ 21)$	$eq_\delta^{Adr,Adr}(20,\ 21)$	$eq_\delta^{E\_Adr,E\_Adr}(20,\ 21)$
20	23	$eq_\delta^{Adr,Adr}(20,\ 23)$	$eq_\delta^{Adr,Adr}(20,\ 23)$	$eq_\delta^{Adr,Adr}(20,\ 23)$
20	25	$eq_\delta^{Adr,Adr}(20,\ 25)$	$eq_\delta^{Adr,Adr}(20,\ 25)$	$eq_\delta^{E\_Adr,E\_Adr}(20,\ 25)$
25	26	$eq_\delta^{Adr,Adr}(25,\ 26)$	$eq_\delta^{Adr,Adr}(25,\ 26)$	$eq_\delta^{JE\_Adr,JE\_Adr}(25,\ 26)$
23	23	$eq_\delta^{Adr,Adr}(23,\ 23)$	$eq_\delta^{Adr,Adr}(23,\ 23)$	$eq_\delta^{J\_Adr,J\_Adr}(23,\ 23)$

9   Sketch of Proofs for Facts and Theorems

Our definitions are provide recursive constructions. Therefore, most Facts and Theorems can be proved by structural recursion.

Fact 2 *Let $S = (C, isa, att, meth)$ be a schema. The subtype ordering $\preceq_{T_C}$ is not necessarily simple even if the inheritance relation isa is simple.*

Proof by counter example: If isa is defined by $c_1 \leq_{isa} c'_1$, $c_2 \leq_{isa} c'_2$ then isa is simple. However, $c_1 \otimes c_2 \preceq_{T_C} c'_1 \otimes c_2$, $c_1 \otimes c_2 \preceq_{T_C} c_1 \otimes c'_2$ but $c'_1 \otimes c_2$ and $c_1 \otimes c'_2$ are not comparable in $(T_C, \preceq_{T_C})$. That is, $\preceq_{T_C}$ is not simple. ∎

Fact 5 *If inheritance is simple then att satisfies the minimum condition. However, meth may not satisfy the minimum condition even if inheritance is simple.*

Proof : In a simple inheritance $super^*(c, a)$ is a finite chain, so it has a minimum. To complete the proof consider a schema with two classes c and c' such that $c \leq_{isa} c'$, $m : c' \to t'$ is the only method in c and $m : c \to t$ is the only method in c'. Obviously, inheritance is simple but $min(super^*(c',))$ doesn't exist. Indeed, $super^*(c', m) = \{(c, c'), (c', c)\}$, $(c, c') \npreceq_{C \times T_C^*} (c', c)$ and $(c', c) \npreceq_{C \times T_C^*} (c, c')$. ∎

Theorem 1 *If $S = (C, isa, att, meth)$ is a hierarchically consistent schema then $\overline{S} = (C, isa, \overline{att}, \overline{meth})$ is also a hierarchically consistent schema that we call the closure of S. Moreover, $(\overline{\overline{S}}) = \overline{S}$.*

Proof of covariance for $\overline{att}$: We have to prove that If $\overline{att}(c_1, a) = t_1$ and $\overline{att}(c_2, a) = t_2$ and $c_1 \leq_{isa} c_2$ then $t_1 \preceq_{T_C} t_2$. By resolution procedure there are c_1' and c_2' such that $ResA(a, c_1) = c_1' = min(super^*(c_1, a))$, $ResA(a, c_2) = c_2' = min(super^*(c_2, a))$, $att(c_1', a) = t_1$ and $att(c_2', a) = t_2$. But $c_1 \leq_{isa} c_2$, therefore $super^*(c_2, a) \subseteq super^*(c_1, a)$ and consequently $c_1' \leq_{isa} c_2'$. Since S satisfies the covariant condition we have $t_1 \preceq_{T_C} t_2$, by Fact 4. The proof of covariance for $\overline{meth}$ is similar.

Proof of minimality for $\overline{att}$: Let $\overline{super}^*(a, c) = \{c' \mid c \leq_{isa} c' \text{ and } \exists\, \overline{tatt}(c', a) = t\}$. We have to prove that if $\overline{super}^*(c, a) \neq \emptyset$ then it has a minimum. But, $\overline{super}^*(a, c) \neq \emptyset$ means that a is an explicit or a derived attribute in c, that is $c \in \overline{super}^*(a, c)$. Thus, $min(\overline{super}^*(a, c)) = c$. The proof of satisfaction of minimal condition for $\overline{meth}$ is similar. ∎

Theorem 2 *For any classification function γ and all types t and t', if $t \preceq_{T_C^u} t'$ then $[\![t]\!]^\gamma \subseteq [\![t']\!]^\gamma$.*

Proof: If $t \leq_{isa} t'$ then $[\![t]\!]^\gamma = [\![t']\!]^\gamma$ when t and t' are class names, or basic types. The general case is proved obviously by structural recursion. ∎

Theorem 3 *For every type t and every $v : t$, $mst(v : t)$ exists and is unique, moreover $mst(v : t) \preceq_{T_C} t$.*

Proof Existence and uniqueness are direct results of the definition of mst, by structural recursion. Let us prove $mst(v : t) \preceq_{T_C} t$.

If t is *unit* or a basic type then $mst(v : t) = t$, and if t is a class name c then $mst(v : t)$ is the subclass of c in which v is classified, that is $mst(v : t) \preceq_{T_C} t$.

If t is a product type $t_1 \otimes \cdots \otimes t_n$ and $v = (v_1, \cdots, v_n)$ then by Definition 12, $mst(v : t) = t_1' \otimes \cdots \otimes t_n'$ where t_i' is either t_i or $mst(v_i : t_i)$. In both cases $mst(v_i : t_i) \preceq_{T_C} t_i$, and $mst(v : t) \preceq_{T_C} t$.

Let $t = set(t_1)$, $v = \{v_1, \cdots, v_n\}$, and assume $t' = lub\{mst(v_1 : t_1), \cdots, mst(v_n : t_1)\}$ exists. By induction hypothesis $mst(v_i : t_1) \preceq_{T_C} t_1$ for all $i = 1, \cdots, n$, that is t_1 is an upper bound of $\{mst(v_1 : t_1), \cdots, mst(v_n : t_1)\}$. This implies that $t' \leq t_1$, therefore $mst(v : t) = set(t') \leq set(t_1) = t$. This complete the proof. ∎

For the proofs of Fact 7 and Fact 8 see pages 116-118 of [30]

Graph-Theoretical Methods in Computer Vision

Ali Shokoufandeh[1] and Sven Dickinson[2]

[1] Department of Mathematics and Computer Science
Drexel University
Philadelphia, PA
USA
ashokoufandeh@mcs.drexel.edu

[2] Department of Computer Science and Center for Cognitive Science
Rutgers University
New Brunswick, NJ
USA
sven@cs.rutgers.edu

Abstract. The management of large databases of hierarchical (e.g., multi-scale or multilevel) image features is a common problem in object recognition. Such structures are often represented as trees or directed acyclic graphs (DAGs), where nodes represent image feature abstractions and arcs represent spatial relations, mappings across resolution levels, component parts, etc. Object recognition consists of two processes: *indexing* and *verification*. In the indexing process, a collection of one or more extracted image features belonging to an object is used to select, from a large database of object models, a small set of candidates likely to contain the object. Given this relatively small set of candidates, a verification, or matching procedure is used to select the most promising candidate. Such matching problems can be formulated as *largest isomorphic subgraph* or *largest isomorphic subtree* problems, for which a wealth of literature exists in the graph algorithms community. However, the nature of the vision instantiation of this problem often precludes the direct application of these methods. Due to occlusion and noise, no significant isomorphisms may exists between two graphs or trees. In this paper, we review our application of spectral encoding of a graph for indexing to large database of image features represented as DAGs. We will also review a more general class of matching methods, called *bipartite matching*, to two problems in object recognition.

1 Introduction

The management of large databases of hierarchical (e.g., multi-scale or multi-level) image features is a common problem in object recognition. Such structures are often represented as trees or DAGs, where nodes represent image feature abstractions and arcs represent spatial relations, mappings across resolution levels, component parts, etc. Object recognition consists of two processes: indexing and verification. In the indexing process, a collection of one or more extracted image features belonging to an object is used to select, from a large database of

G.B. Khosrovshahi et al. (Eds.): Theoretical Aspects of Computer Science, LNCS 2292, pp. 148–174, 2002.
© Springer-Verlag Berlin Heidelberg 2002

object models, a small set of candidates likely to contain the object. Given this relatively small set of candidates, a verification, or matching procedure is used to select the most promising candidate. The requirements of matching include computing a correspondence between nodes in an image structure and nodes in a model structure, as well as computing an overall measure of distance (or, alternatively, similarity) between the two structures. Such matching problems can be formulated as *largest isomorphic subgraph* or *largest isomorphic subtree* problems, for which a wealth of literature exists in the graph algorithms community. However, the nature of the vision instantiation of this problem often precludes the direct application of these methods. Due to occlusion and noise, no significant isomorphisms may exists between two graphs or trees. Yet, at some level of abstraction, the two structures (or two of their substructures) may be quite similar.

In this paper, we review our application of spectral encoding of a graph for indexing to large database of image features represented as DAG. Our indexing mechanism maps the topological structure of a DAG into a low-dimensional vector space, based on eigenvalue characterization of its adjacency matrix. Invariant to any re-ordering of the DAG's branches, the vector provides an invariant signature of the shape's coarse topological structure. Furthermore, we can efficiently index into a database of topological signatures to retrieve model objects having similar topology. In a set of experiments, we show that the indexing mechanism is very effective in selecting a small set of model candidates that contain the correct object.

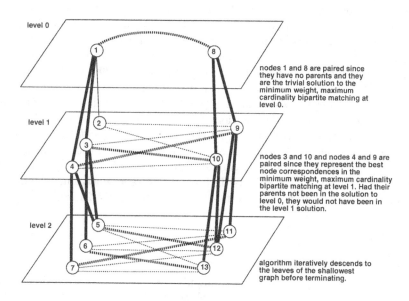

level 0

nodes 1 and 8 are paired since they have no parents and they are the trivial solution to the minimum weight, maximum cardinality bipartite matching at level 0.

level 1

nodes 3 and 10 and nodes 4 and 9 are paired since they represent the best node correspondences in the minimum weight, maximum cardinality bipartite matching at level 1. Had their parents not been in the solution to level 0, they would not have been in the level 1 solution.

level 2

algorithm iteratively descends to the leaves of the shallowest graph before terminating.

Fig. 1. Bipartite Matching

We will also review our application of a more general class of matching methods, called *bipartite matching*, to problems in object recognition [25,26,28,30,31]. As shown in Fig. 1, given two graphs (or trees) G_1 and G_2, $H(G_1, G_2, E)$ is a weighted bipartite graph with weight matrix $W = [w_{u,v}]$ of size $|G_1| \times |G_2|$, for all edges of the form $(u, v) \in E$, $u \in G_1$, $v \in G_2$, and (u, v) has an associated weight $= w_{u,v}$. Solving the maximum cardinality minimum weight matching in H solves an optimization problem which tries to minimize total edge weight on one hand while trying to maximize the total number of edges in the solution set on the other hand. The time complexity for finding such a matching in a weighted bipartite graph with n vertices is $O(n^2 \sqrt{n} \log \log n)$ time, using the scaling algorithm of Gabow, Gomans and Williamson [11].

We will apply this framework for solving two object recognition problems, one involving DAGs and one involving rooted trees. Each algorithm will, as an integral step, compute the maximum cardinality, minimum weight matching in a bipartite graph. Furthermore, each algorithm, in turn, takes a different approach to preserving hierarchical order in the solution. We describe each algorithm in detail and evaluate its performance on sets of real images.

2   Two Object Recognition Domains

2.1   The Saliency Map Graph

Our first image representation is a multi-scale view-based description of 3-D objects that, on one hand, avoids the need for complex feature extraction, such as lines, curves, or regions, while on the other hand, provides the locality of representation necessary to support occluded object recognition as well as invariance to minor changes in both illumination and shape. In computing a representation for a 2-D image, a multi-scale wavelet transform is applied to the image, resulting in a hierarchical map that captures salient regions at their appropriate scales of resolution. Each such region maps to a node in a DAG, in which an arc is directed from a coarser scale region to a finer scale region if the center of the finer scale's region falls within the interior of the coarser scale's region. The resulting hierarchical graph structure, called the *saliency map graph* (SMG), encodes both the topological and geometrical information found in the saliency map. An example of an image and its corresponding saliency map graph are shown in Fig.s 2(a) and (b), respectively. Details of the representation, including its computation and invariance properties, can be found in [25,26,28].

2.2   Shock Trees

Our second image representation describes the generic shape of a 2-D object, and is based on a coloring of the shocks (singularities) of a curve evolution process acting on simple closed curves in the plane [15]. Intuitively, the taxonomy of shocks consists of four distinct types: the radius function along the medial axis varies monotonically at a 1, achieves a strict local minimum at a 2, is constant at

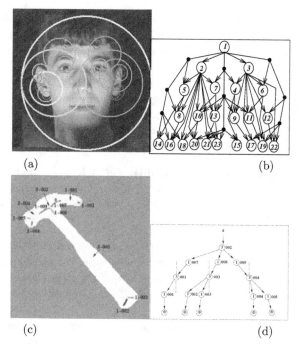

Fig. 2. Two Object Recognition Domains: (a) example image; (b) saliency map graph corresponding to image in (a); (c) example silhouette with computed shocks; (d) shock tree corresponding to silhouette in (c).

a 3 and achieves a strict local maximum at a 4. We have recently abstracted this system of shocks into a *shock graph* where vertices are labelled by their shock types, and the shock formation times direct the edges. The space of such shock graphs is completely characterized by a small number of rules, which in turn permits the reduction of each graph to a *unique rooted tree* [30,31]. Figure 2(c) and (d) show the the 2-D silhouette of a hammer and its corresponding shock tree, respectively.

3   Indexing Mechanism for Directed Acyclic Graphs

3.1   An Eigenvalue Characterization of a DAG

To describe the topology of a DAG, we turn to the domain of eigenspaces of graphs, first noting that any graph can be represented as a $\{-1, 0, 1\}$ adjacency matrix, with 1's (and -1's) indicating directed edges in the graph (and 0's on the diagonal). The eigenvalues of a graph's (or DAG's) adjacency matrix encode important structural properties of the graph (or DAG). Furthermore, the eigenvalues of a symmetric matrix A are invariant to any orthonormal transformation of the form $P^t A P$. Since a permutation matrix is orthonormal, the eigenvalues of a DAG are invariant to any consistent re-ordering of the DAG's branches. However, before we can exploit a DAG's eigenvalues for matching purposes, we must establish their stability under minor topological perturbation, due to noise, occlusion, or deformation.

We begin with the case in which the image DAG is formed by either adding a new root to the model DAG, adding one or more subgraphs at leaf nodes of the model DAG, or deleting one or more entire model subgraphs. In this case, the model DAG is a subgraph of the query DAG, or vice versa. The following theorem relates the eigenvalues of two such DAGs:

Theorem 1 (see Cvetković et al. [6]). *Let A be a symmetric[1] matrix with eigenvalues $\lambda_1 \geq \lambda_2 \geq \ldots \geq \lambda_n$ and let B be one of its principal[2] sub-matrices. If the eigenvalues of B are $\nu_1 \geq \nu_2 \geq \ldots \geq \nu_m$, then $\lambda_{n-m+i} \leq \nu_i \leq \lambda_i (i = 1, \ldots, m)$.*

This important theorem, called the *Interlacing Theorem*, implies that as A and B become less similar (in the sense that one is a smaller subgraph of the other), their eigenvalues become proportionately less similar (in the sense that the intervals that contain them increase in size, allowing corresponding eigenvalues to drift apart).

The other case we need to consider consists of a query DAG formed by adding to or removing from the model DAG, a small subset of internal (i.e., non-leaf) nodes. The upper bounds on the two largest eigenvalues ($\lambda_1(T)$ and $\lambda_2(T)$) of any DAG, T, with n nodes and maximum degree $\Delta(T)$ are $\lambda_1(T) \leq \sqrt{n-1}$ and $\lambda_2(T) \leq \sqrt{(n-3)/2}$, respectively (Neumaier, 1982 [18]). The lower bounds on these two eigenvalues are $\lambda_1(T) \geq \sqrt{\Delta(T)}$ (Nosal, 1970 [19]) and $\lambda_1(T)\lambda_2(T) \geq \frac{2n-2}{n-2}$ (Cvetković, 1971 [5]). Therefore, the addition or removal of a small subset of internal nodes will result in a small change in the upper and lower bounds on these two eigenvalues. As we shall next, our topological description exploits the largest eigenvalues of a DAG's adjacency matrix. Since these largest eigenvalues are stable under minor perturbation of the DAG's internal node structure, so too is our topological description.

We now seek a compact representation of the DAG's topology based on the eigenvalues of its adjacency matrix. We could, for example, define a vector to be the sorted eigenvalues of a DAG. The resulting index could be used to retrieve nearest neighbors in a model DAG database having similar topology. There is a problem with this approach. For large DAGs, the dimensionality of the signature would be prohibitively large. To solve this problem, this description will be based on eigenvalue sums rather than on the eigenvalues themselves.

Specifically, let T be a DAG whose maximum branching factor is $\Delta(T)$, and let the subgraphs of its root be $T_1, T_2, \ldots, T_S$. For each subgraph, T_i, whose root degree is $\delta(T_i)$, compute the eigenvalues of T_i's sub-matrix, sort the eigenvalues

---

[1] The original theorem is stated for Hermitian matrices, of which symmetric matrices are a subclass.

[2] A principal sub-matrix of a graph's adjacency matrix is formed by selecting the rows and columns that correspond to a subset of the graph's nodes.

in decreasing order by absolute value, and let S_i be the sum of the $\delta(T_i) - 1$ largest absolute values. As shown in Fig. 3, the sorted S_i's become the components of a $\Delta(T)$-dimensional vector assigned to the DAG's root. If the number of S_i's is less than $\Delta(T)$, then the vector is padded with zeroes. We can recursively repeat this procedure, assigning a vector to the root of each subgraph in the DAG for reasons that will become clear in the next section.

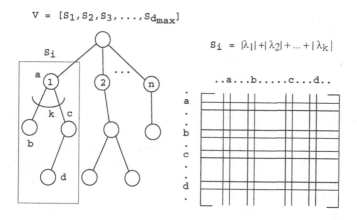

d_{max} = max degree of any scene or model node

k = degree of a given subtree's root

Fig. 3. Computing a Topological Signature of a DAG

Although the eigenvalue sums are invariant to any consistent re-ordering of the DAG's branches, we have given up some uniqueness (due to the summing operation) in order to reduce dimensionality. We could have elevated only the largest eigenvalue from each subgraph (non-unique but less ambiguous), but this would be less representative of the subgraph's structure. We choose the $\delta(T_i) - 1$-largest eigenvalues for two reasons: 1) the largest eigenvalues are more informative of subgraph structure, 2) by summing $\delta(T_i) - 1$ elements, we effectively normalize the sum according to the local complexity of the subgraph root.

To efficiently compute the sub-matrix eigenvalue sums, we turn to the domain of semidefinite programming. A symmetric $n \times n$ matrix A with real entries is said to be positive semidefinite, denoted as $A \succeq 0$, if for all vectors $x \in R^n$, $x^t A x \geq 0$, or equivalently, all its eigenvalues are non-negative. We say that $U \succeq V$ if the matrix $U - V$ is positive semidefinite. For any two matrices U and V having the same dimensions, we define $U \bullet V$ as their inner product, i.e.,

$U \bullet V = \sum_i \sum_j U_{i,j} V_{i,j}$. For any square matrix U, we define $\text{trace}(U) = \sum_i U_{i,i}$. Let I denote the identity matrix having suitable dimensions. The following result, due to Overton and Womersley [20], characterizes the sum of the first k largest eigenvalues of a symmetric matrix in the form of a semidefinite convex programming problem:

Theorem 2 (Overton and Womersley [20]). *For the sum of the first k eigenvalues of a symmetric matrix A, the following semidefinite programming characterization holds:*

$$\lambda_1(A) + \ldots + \lambda_k(A) = \max A \bullet U$$
$$\text{s.t.} \quad \text{trace}(U) = k \qquad (1)$$
$$0 \preceq U \preceq I,$$

The elegance of Theorem (2) lies in the fact that the equivalent semidefinite programming problem can be solved, for any desired accuracy ϵ, in time polynomial in $O(n\sqrt{n}L)$ and $\log \frac{1}{\epsilon}$, where L is an upper bound on the size of the optimal solution, using a variant of the Interior Point method proposed by Alizadeh [1]. In effect, the complexity of directly computing the eigenvalue sums is a significant improvement over the $O(n^3)$ time required to compute the individual eigenvalues, sort them, and sum them.

3.2   A Database for Model DAGs

Our eigenvalue characterization of a DAG's topology suggests that a model DAG's topological structure can be represented as a vector in δ-dimensional space, where δ is an upper bound on the degree of any vertex of any image or model DAG. If we could assume that an image DAG represents a properly segmented, unoccluded object, then the vector of eigenvalue sums, call it the *topological signature vector* (or TSV), computed at the image DAG's root could be compared with those topological signature vectors representing the roots of the model DAGs. The vector distance between the image DAG's root TSV and a model DAG's root TSV would be inversely proportional to the topological similarity of their respective DAGs: recall that finding two subgraphs with "close" eigenvalue sums represents an approximation to finding the largest isomorphic subgraph.

Unfortunately, this simple framework cannot support either cluttered scenes or segmentation errors, both of which result in the addition or removal of DAG structure. In either case, altering the structure of the DAG will affect the TSV's computed at its nodes. The signatures corresponding to those subgraphs that survive the occlusion will not change. However, the signature of a node having one or more subgraphs which have undergone any perturbation will change which, in turn, will affect the signatures of any of its ancestor nodes, including the root. We therefore cannot rely on indexing solely with the root's signature. Instead, we will take advantage of the local subgraphs that survive the occlusion.

We can accommodate such perturbations through a local indexing framework analogous to that used in a number of geometric hashing methods, e.g., [17,10]. Rather than storing a model DAG's root signature, we will store the signatures of *each* node in the model DAG, along with a pointer to the object model containing that node as well as a pointer to the corresponding node in the model DAG (allowing access to node label information). Since a given model subgraph can be shared by other model DAGs, a given signature (or location in δ-dimensional space) will point to a list of (model object, model node) ordered pairs. At runtime, the signature at each node in the image DAG becomes a separate index, with each nearby candidate in the database "voting" for one or more (model object, model node) pairs. To quickly retrieve these nearby candidates, we will pre-compute the sorted pairwise distances between every signature in the database and every other signature in the database.

3.3   An Efficient Indexing Mechanism

We achieve efficient indexing through a δ-dimensional Voronoi decomposition $P(B)$ of the model space $V(B)$. For a given image TSV, the Voronoi decomposition will allow us to find the nearest model TSV in expected $O(\log^\delta(kn))$ time for fixed δ [22]. From the ranked list of neighbors L computed for each model TSV, either the ℓ nearest neighbors or all neighbors within a radius r can be computed in constant time (assuming fixed ℓ or r). To construct the Voronoi database, we partition R^δ into regions, so that for each vector $v \in V(B)$, the region $P(v)$ will denote the set of points in R^δ which are closer to v than any other vector in $V(B)$ with respect to a metric norm, such as $d(.,.)$. Such a partition is well-defined. The complexity of the Voronoi decomposition is $O((kn)^{\lfloor(\delta+1)/2\rfloor+1}) + O((kn)^{\lfloor(\delta+1)/2\rfloor} \log(kn))$ ([22]), although this is a cost incurred at preprocessing time.

The process of selecting candidates at runtime is shown in Fig. 4. Let H be an image DAG, and let F_H and $V(F_H)$ be defined as before, and let $d(u,v) = ||v - u||_2$. For each TSV $h \in V(F_H)$, we will find the region $P(v)$ (and corresponding vector v) in $P(B)$, in which h resides. Using the list $L(v)$, we will find the set of model TSV's $\{u_1,..,u_\ell\}$ such that $d(h,v) + d(v,u_i) \leq r$. Clearly, since the metric norm $d(.,.)$ satisfies the triangle inequality, the set $\{v\} \cup \{u_1,..,u_\ell\}$ is a subset of the TSV's whose distance from h is less than r. Each node in the image DAG therefore leads to a number of (model object, model node) candidate votes. In the next section, we discuss the weighting of these votes, along with the combination of the evidence over all nodes in the image DAG.

4   Matching Two Saliency Map Graphs

Given the SMG computed for an input image to be recognized and a SMG computed for a given model object image (view), we propose two methods for computing their similarity. In the first method, we compare only the topological or structural similarity of the graphs, a weaker distance measure designed to

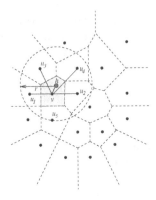

Fig. 4. Selecting the Candidate Model Objects

support limited object deformation invariance. In the second method, we take advantage of the geometrical information encoded in an SMG and strengthen the similarity measure to ensure geometric consistency, a stronger distance measure designed to support subclass or instance matching. Each method is based on formulating the problem as a maximum cardinality minimum weight matching in a bipartite graph.

4.1   Problem Formulation

Two graphs $G = (V, E)$ and $G' = (V', E')$ are said to be isomorphic if there exists a bijective mapping $f : V \to V'$ satisfying, for all $x, y \in V$ $(x, y) \in E \Leftrightarrow (f(x), f(y)) \in E'$. To compute the similarity of two SMG's, we consider a graph isomorphism problem, which we will call the *SMG similarity problem*: Given two SMG's $G_1 = (V_1, E_1)$ and $G_2 = (V_2, E_2)$ and a partial mapping from $f : V_1 \to V_2$, let $\mathcal{E}$ be a real-valued error function defined on the set of all partial mappings. Our error function, $\mathcal{E}$, incorporates two components with respect to any partial mapping: 1) we would like to reward corresponding nodes which are similar in terms of their topology, geometry, and salience; and 2) we would like to penalize a set of correspondences the more they exclude nodes from the model. Specifically,

$$\mathcal{E}(f) = \varepsilon \sum_{u \in V_1, v \in V_2} M_{u,v}\, \omega(u, v)\, |s(u) - s(v)| + (1 - \varepsilon) \sum_{u \in V_1, f(u) = \emptyset} s(u) \quad (2)$$

where $\varepsilon = |\mathbf{1}^t M(f)\mathbf{1}|/(|V_1| + |V_2|)$ represents the fraction of matched vertices ($\mathbf{1}$ denotes the identity vector), $f(.) = \emptyset$ for unmatched vertices, and $s(.)$ represents region saliency. For the SMG topological similarity, Sect. 4.2, $\omega(., .)$ is always one, while for the SMG geometrical similarity, Sect. 4.3, it denotes the Euclidean distance between the regions.[3] A more detailed discussion of the error function is provided in [28]. We say that a partial mapping f is feasible if $f(x) = y$ implies that there are parents p_x of x and p_y of y, such that $f(p_x) = p_y$. Our goal is therefore to find a feasible mapping f which minimizes $\mathcal{E}(f)$.

4.2   A Matching Algorithm Based on Topological Similarity

In this section, we describe an algorithm which finds an approximate solution to the SMG similarity problem. The focus of the algorithm is to find a minimum weight matching between vertices of G_1 and G_2 which lie in the same level. Our algorithm starts with the vertices at level 1. Let A_1 and B_1 be the set of vertices at level 1 in G_1 and G_2, respectively. We construct a complete weighted bipartite graph $G(A_1, B_1, E)$ with a weight function defined for edge (u, v) ($u \in A_1$ and $v \in B_1$) as $w(u, v) = |s(v) - s(u)|$.[4] Next, we find a maximum cardinality, minimum weight matching M_1 in G using [8]. All the matched vertices are mapped to each other; that is, we define $f(x) = y$ if (x, y) is a matching edge in M_1.

The remainder of the algorithm proceeds in phases as follows, as shown in Fig. 5. In phase i, the algorithm considers the vertices of level i. Let A_i and B_i be the set of vertices of level i in G_1 and G_2, respectively. Construct a weighted bipartite graph $G(A_i, B_i, E)$ as follows: (v, u) is an edge of G if either of the following is true: (1) Both u and v do not have any parent in G_1 and G_2, respectively, or (2) They have at least one matched parent of depth less than i; that is, there is a parent p_u of u and p_v of v such that $(p_u, p_v) \in M_j$ for some $j < i$. We define the weight of the edge (u, v) to be $|s(u) - s(v)|$. The algorithm finds a maximum cardinality, minimum weight matching in G and proceeds to the next phase.

The above algorithm terminates after ℓ phases, where ℓ is the minimum number of scales in the saliency maps (or SMG's) of two graphs. The partial mapping M of SMG's can be simply computed as the union of all M_i values for $i = 1, \ldots, \ell$. Finally, using the error measure defined in [28], we compute the error of the partial mapping M. Each phase of the algorithm requires simple operations with the time to complete each phase being dominated by the time to compute a minimum weight matching in a bipartite graph. As mentioned in Sect. 1, the time complexity for finding such a matching in a weighted bipartite graph with n vertices is $O(n^2\sqrt{n} \log \log n)$ time, using the scaling algorithm of Gabow, Gomans and Williamson [11]. The entire procedure, as currently formulated, requires $O(\ell n^2 \sqrt{n} \log \log n)$ steps.

---

[3] For perfect similarity $\mathcal{E}(f) = 0$, while $\mathcal{E}(f)$ will be $\sum_{u \in V_1} s(u)$ if there is no match.

[4] $G(A, B, E)$ is a weighted bipartite graph with weight matrix $W = [w_{ij}]$ of size $|A| \times |B|$ if, for all edges of the form $(i, j) \in E$, $i \in A$, $j \in B$, and (i, j) has an associated weight $= w_{i,j}$.

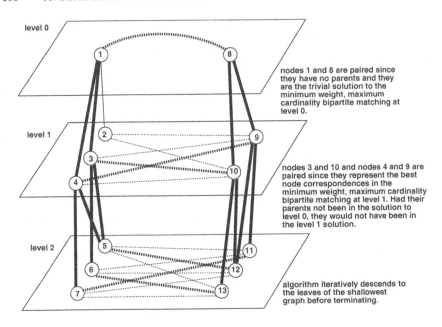

level 0

nodes 1 and 8 are paired since
they have no parents and they
are the trivial solution to the
minimum weight, maximum
cardinality bipartite matching at
level 0.

level 1

nodes 3 and 10 and nodes 4 and 9 are
paired since they represent the best
node correspondences in the
minimum weight, maximum cardinality
bipartite matching at level 1. Had their
parents not been in the solution to
level 0, they would not have been in
the level 1 solution.

level 2

algorithm iteratively descends to
the leaves of the shallowest
graph before terminating.

Fig. 5. Illustration of the SMGBM Algorithm (see text for explanation).

4.3   A Matching Algorithm Based on Geometric Similarity

The SMGBM similarity measure captured the structural similarity between two
SMG's in terms of branching factor and node saliency similarity; no geometric
information encoded in the SMG was exploited. In this section, we describe a
second similarity measure, called SMG Similarity using an Affine Transformation
(SMGAT), that includes the geometric properties (e.g., relative position and
orientation) of the saliency regions.

Given $G_1 = (V_1, E_1)$ and $G_2 = (V_2, E_2)$, we first assume, without loss of
generality, that $|V_1| \leq |V_2|$. First, as shown in Fig. 6, the algorithm will hypoth-
esize a correspondence between three regions of G_1, say (r_1, r_2, r_3), and three
regions (r'_1, r'_2, r'_3) of G_2. The mapping $\{(r_1 \to r'_1), (r_2 \to r'_2), (r_3 \to r'_3)\}$ will be
considered as a basis for alignment if the following conditions are satisfied:

- r_i and r'_i have the same level in the SMG's, for all $i \in \{1, \dots, \ell\}$.
- $(r_i, r_j) \in E_1$ if and only if $(r'_i, r'_j) \in E_2$, for all $i, j \in \{1, \dots, \ell\}$, which
  implies that selected regions should have the same adjacency structure in
  their respective SMG's.

Once regions (r_1, r_2, r_3) and (r'_1, r'_2, r'_3) have been selected, we solve for the
affine transformation (A, b), that aligns the corresponding region triples by solving the following system of linear equalities:

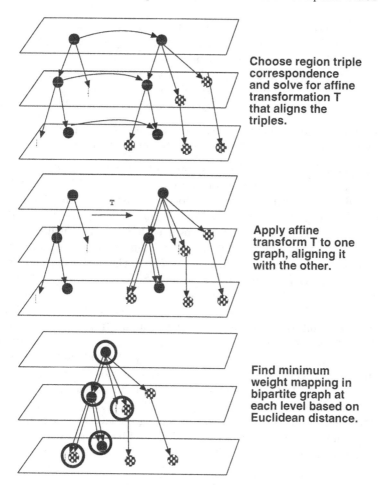

Choose region triple correspondence and solve for affine transformation T that aligns the triples.

Apply affine transform T to one graph, aligning it with the other.

Find minimum weight mapping in bipartite graph at each level based on Euclidean distance.

Fig. 6. Illustration of the SMGAT Algorithm (see text for explanation)

$$
\begin{bmatrix}
x_{r_1} & y_{r_1} & 1 & 0 & 0 & 0 \\
x_{r_2} & y_{r_2} & 1 & 0 & 0 & 0 \\
x_{r_3} & y_{r_3} & 1 & 0 & 0 & 0 \\
0 & 0 & 0 & x_{r_1} & y_{r_1} & 1 \\
0 & 0 & 0 & x_{r_2} & y_{r_2} & 1 \\
0 & 0 & 0 & x_{r_3} & y_{r_3} & 1
\end{bmatrix}
\begin{bmatrix}
a_{11} \\
a_{12} \\
b_1 \\
a_{21} \\
a_{22} \\
b_2
\end{bmatrix}
=
\begin{bmatrix}
x_{r'_1} \\
x_{r'_2} \\
x_{r'_3} \\
y_{r'_1} \\
y_{r'_2} \\
y_{r'_3}
\end{bmatrix}.
\tag{3}
$$

The affine transformation (A, b) will be applied to all regions in G_1 to form a new graph G'. Next, a procedure similar to the minimum weight matching, used in the SMGBM is applied to the regions in graphs G' and G_2. Instead of matching regions which have maximum similarity in terms of saliency, we match regions which have minimum Euclidean distance from each other. Given two regions u

and v, the distance between them can be defined as the L_2 norm of the distance between their centers, denoted by $d(u,v) = \sqrt{(x_u - x_v)^2 + (y_u - y_v)^2}$. In a series of steps, SMGAT constructs weighted bipartite graphs $\mathcal{G}_i = (R_i, R'_i, E_i)$ for each level i of the two SMG's, where R_i and R'_i represent the set of vertices of G' and G_2 at the i-th level, respectively. The constraints for having an edge in E_i are the same as SMGBM: (u,v) is an edge in $\mathcal{G}_i$ if either of the followings holds:

- Both u and v do not have any parents in G' and G_2, respectively.
- They have at least one matched parent of depth less than i.

The corresponding edge will have weight equal to $w(u,v) = d(u,v)$. A maximum cardinality, minimum weight bipartite matching M_i will be found for each level $\mathcal{G}_i$, and the partial mapping $f_{(A,b)}$ for the affine transformation (A, b) will be formed as the union of all M_i's. Finally, the error of this partial mapping $\mathcal{E}(f_{(A,b)})$ will be computed as the sum over each E_i of the Euclidean distance separating E_i's nodes weighted by the nodes' difference in saliency. Once the total error is computed, the algorithm proceeds to the next valid pair of region triples. Among all valid affine transformations, SMGAT chooses that one which minimizes the error of the partial mapping.

In terms of algorithmic complexity, solving for the affine transformation (3) takes only constant time, while applying the affine transformation to G_1 to form G' is $O(\max(|V_1|, |E_1|))$. The execution time for each hypothesized pair of region triples is dominated by the complexity of establishing the bipartite matching between G_2 and G', which is $O(\ell n^2 \sqrt{n} \log \log n)$, for SMG's with n vertices and ℓ scales. In the worst case, i.e., when both saliency map graphs have only one level, there are $O(n^6)$ pairs of triples. However, in practice, the vertices of an SMG are more uniformly distributed among the levels of the graph, greatly reducing the number of possible correspondences of base triples. For a discussion of how the complexity of the bipartite matching step can be reduced, see [27].

5   Matching Two Shock Trees

5.1   Problem Formulation

Given two shock graphs, one representing an object in the scene (V_2) and one representing a database object (V_1), we seek a method for computing their similarity. Unfortunately, due to occlusion and clutter, the shock graph representing the scene object may, in fact, be embedded in a larger shock graph representing the entire scene. Thus we have a *largest subgraph isomorphism* problem, stated as follows: Given two graphs $G = (V_1, E_1)$ and $H = (V_2, E_2)$, find the maximum integer k, such that there exists two subsets of cardinality k, $E'_1 \subseteq E_1$ and $E'_2 \subseteq E_2$, and the induced subgraphs (not necessarily connected) $G' = (V_1, E'_1)$ and $H' = (V_2, E'_2)$ are isomorphic [12]. Further, since our shock graphs are labeled graphs, consistency between node labels must be enforced in the isomorphism.

The largest subgraph isomorphism problem, can be formulated as a $\{0, 1\}$ integer optimization problem. The optimal solution is a $\{0, 1\}$ bijective mapping matrix M, which defines the correspondence between the vertices of the two graphs G and H, and which minimizes an appropriately defined distance measure between corresponding edge and/or node labels in the two graphs.

We seek the matrix M, the global optimizer of the following [16,7]:

$$\min -\frac{1}{2} \sum_{u \in V_1} \sum_{v \in V_2} M(u,v)\|u,v\|$$
$$\text{s.t.} \quad \sum_{u' \in V_2} M(u,u') \leq 1, \ \forall u \in V_1$$
$$\sum_{v \in V_1} M(v,v') \leq 1, \ \forall v' \in V_2 \tag{4}$$
$$M(x,y) \in \{0,1\}, \forall x \in V_1, \ y \in V_2$$

where $\|.\|$ is a measure of the similarity between the labels of corresponding nodes in the two shock graphs (see Sect. 5.2).

The above minimization problem is known to be NP-hard for general graphs [12], however, polynomial time algorithms exist for the special case of finite rooted trees with no vertex labels. Matula and Edmonds [9] describe one such technique, involving the solution of $2n_1 n_2$ network flow problems, where n_1 and n_2 represent the number of vertices in the two graphs. The complexity was further reduced by Reyner [24] to $O(n_1^{1.5} n_2)$ (assuming $n_1 \geq n_2$), through a reduction to the bipartite matching algorithm of Hopcraft and Karp [14]. Since we can transform any shock graph into a unique rooted shock tree [30,31], we can pursue a polynomial time solution to our problem. However, as mentioned in Sect. 1, the introduction of noise (spurious addition/deletion of nodes) and/or occlusion may prevent the existence of large isomorphic subtrees. We therefore need a matching algorithm that can find isomorphic subtrees under these conditions. To accomplish this, we have developed a topological representation for trees that is invariant to minor perturbations in structure.

5.2   The Distance between Two Vertices

The eigenvalue characterization introduced in the previous section applies to the problem of determining the topological similarity between two shock trees. This, roughly speaking, defines an equivalence class of objects having the same structure but whose parts may have different qualitative or quantitative shape. For example, a broad range of 4-legged animals will have topologically similar shock trees.

This geometry is encoded by information contained in each vertex of the shock tree. Recall from Sect. 2.2 that $\tilde{1}$'s and $\tilde{3}$'s represent curve segments of shocks. We choose not to explicitly assign label types 2 and 4, because each may be viewed as a limit case when the number of shocks in a $\tilde{3}$, in the appropriate context, approaches 1 (see Sect. 2.2). Each shock in a segment is further labeled by its position, its time of formation (radius of the skeleton), and its direction of flow (or orientation in the case of $\tilde{3}$'s), all obtained from the shock detection algorithm [29]. In order to measure the similarity between two vertices u

and v, we interpolate a low dimensional curve through their respective shock trajectories, and assign a cost $C(u,v)$ to an affine transformation that aligns one interpolated curve with the other. Intuitively, a low cost is assigned if the underlying structures are scaled or rotated versions of one another (details can be found in [30,31]).

5.3   Algorithm for Matching Two Shock Trees

As stated in Sect. 1, large isomorphic subtrees may not exist between an image shock tree and a model shock tree, due to noise and/or occlusion. A weaker formulation of the problem would be to find the maximum cardinality, minimum weight matching in a bipartite graph spanning the nodes between two shock trees, with edge weights some function of topological distance and geometrical distance. Although the resulting optimization formulation is more general, allowing nodes in one tree to match any nodes in another tree (thereby allowing nodes to match over "noise" nodes), the formulation is weaker since is doesn't enforce hierarchical ordering among nodes. Preserving such ordering is essential, for it makes little sense for a node ordering in one tree to match a reverse ordering in another tree. Unfortunately, we are not aware of a polynomial-time algorithm for solving the bipartite matching problem subject to hierarchical constraints. To achieve a polynomial time approximation, we will embed a bipartite matching procedure into a recursive greedy algorithm that will look for maximally similar subtrees.

Our recursive algorithm for matching the rooted subtrees G and H corresponding to two shock graphs is inspired by the algorithm proposed by Reyner [24]. The algorithm recursively finds matches between vertices, starting at the root of the shock tree, and proceeds down through the subtrees in a depth-first fashion. The notion of a match between vertices incorporates two key terms: the first is a measure of the topological similarity of the subtrees rooted at the vertices (see Sect. 3.1), while the second is a measure of the similarity between the shock geometry encoded at each node (see Sect. 5.2). Unlike a traditional depth-first search which backtracks to the next statically-determined branch, our algorithm effectively recomputes the branches at each node, always choosing the next branch to descend in a best-first manner. One very powerful feature of the algorithm is its ability to match two trees in the presence of noise (random insertions and deletions of nodes in the subtrees).

Before stating our algorithm, some definitions are in order. Let $G = (V_1, E_1)$ and $H = (V_2, E_2)$ be the two shock graphs to be matched, with $|V_1| = n_1$ and $|V_2| = n_2$. Define d to be the maximum degree of any vertex in G and H, i.e., $d = \max(\delta(G), \delta(H))$. For each vertex v, we define $\chi(v) \in R^{d-1}$ as the unique eigen-decomposition vector introduced in Sect. 3.1.[5] Furthermore, for any pair

---

[5] Note that if the maximum degree of a node is d, then excluding the edge from the node's parent, the maximum number of children is $d - 1$. Also note that if $\delta(v) < d$, then then the last $d - \delta(v)$ entries of χ are set to zero to ensure that all χ vectors have the same dimension.

of vertices u and v, let $C(u,v)$ denote the shock distance between u and v, as defined in Sect. 5.2. Finally, let $\Phi(G,H)$ (initially empty) be the set of final node correspondences between G and H representing the solution to our matching problem.

The algorithm begins by forming a $n_1 \times n_2$ matrix $\Pi(G,H)$ whose (u,v)-th entry has the value $C(u,v)\|\chi(u)-\chi(v)\|_2$, assuming that u and v are compatible in terms of their shock order, and has the value ∞ otherwise.[6] Next, we form a bipartite edge weighted graph $\mathcal{G}(V_1, V_2, E_{\mathcal{G}})$ with edge weights from the matrix $\Pi(G,H)$.[7] Using the scaling algorithm of Goemans, Gabow, and Williamson [11], we then find the maximum cardinality, minimum weight matching in $\mathcal{G}$. This results in a list of node correspondences between G and H, called $\mathcal{M}_1$, that can be ranked in decreasing order of similarity.

From $\mathcal{M}_1$, we choose (u_1, v_1) as the pair that has the minimum weight among all the pairs in $\mathcal{M}_1$, i.e., the first pair in $\mathcal{M}_1$. (u_1, v_1) is removed from the list and added to the solution set $\Phi(G,H)$, and the remainder of the list is *discarded*. For the subtrees G_{u_1} and H_{v_1} of G and H, rooted at nodes u_1 and v_1, respectively, we form the matrix $\Pi(G_{u_1}, H_{v_1})$ using the same procedure described above. Once the matrix is formed, we find the matching $\mathcal{M}_2$ in the bipartite graph defined by weight matrix $\Pi(G_{u_1}, H_{v_1})$, yielding another ordered list of node correspondences. The procedure is recursively applied to (u_2, v_2), the edge with minimum weight in $\mathcal{M}_2$, with the remainder of the list discarded.

This recursive process eventually reaches the leaves of the subtrees, forming a list of ordered correspondence lists (or matchings) $\{\mathcal{M}_1, \ldots, \mathcal{M}_k\}$. In backtracking step i, we remove any subtrees from the graphs G_i and H_i whose roots participate in a matching pair in $\Phi(G,H)$ (we enforce a one-to-one correspondence of nodes in the solution set). Then, in a depth-first manner, we first recompute $\mathcal{M}_i$ on the subtrees rooted at u_i and v_i (with solution set nodes removed). As before, we choose the minimum weight matching pair, and recursively descend. Unlike in a traditional depth-first search, we dynamically recompute the branches at each node in the search tree. Processing at a particular node will terminate when either subtree loses all of its nodes to the solution set. We can now state the algorithm more precisely:

procedure isomorphism(G,H)
   $\Phi(G,H) \leftarrow \emptyset$
   $d \leftarrow \max(\delta(G), \delta(H))$
   for $u \in V_G$ compute $\chi(u) \in R^{d-1}$ (Section 3.1)
   for $v \in V_H$ compute $\chi(v) \in R^{d-1}$ (Section 3.1)
   call match$(\text{root}(G),\text{root}(H))$
   return$(\text{cost}(\Phi(G,H)))$
end

---

[6] If either $C(u,v)$ or $\|\chi(u) - \chi(v)\|_2$ is zero, the (u,v)-th entry is the other term.

[7] $G(A,B,E)$ is a weighted bipartite graph with weight matrix $W = [w_{ij}]$ of size $|A| \times |B|$ if, for all edges of the form $(i,j) \in E$, $i \in A$, $j \in B$, and (i,j) has an associated weight $= w_{i,j}$.

procedure match(u,v)
   do
       {
       let $G_u \leftarrow$ rooted subtree of G at u
       let $H_v \leftarrow$ rooted subtree of H at v
       compute $|V_{G_u}| \times |V_{H_v}|$
       weight matrix $\Pi(G_u, H_v)$
       $\mathcal{M} \leftarrow$ max cardinality, minimum weight
          bipartite matching in $\mathcal{G}(V_{G_u}, V_{H_v})$
          with weights from $\Pi(G_u, H_v)$ (see [11])
       $(u',v') \leftarrow$ minimum weight pair in $\mathcal{M}$
       $\Phi(G,H) \leftarrow \Phi(G,H) \cup \{(u',v')\}$
       call match(u',v')
       $G_u \leftarrow G_u - \{x|x \in V_{G_u} \text{ and } (x,w) \in \Phi(G,H)\}$
       $H_v \leftarrow H_v - \{y|y \in V_{H_v} \text{ and } (w,y) \in \Phi(G,H)\}$
       }
   while ($G_u \neq \emptyset$ and $H_v \neq \emptyset$)

In terms of algorithmic complexity, observe that during the depth-first construction of the matching chains, each vertex in G or H will be matched at most once in the forward procedure. Once a vertex is mapped, it will never participate in another mapping again. The total time complexity of constructing the matching chains is therefore bounded by $O(n^2\sqrt{n \log\log n})$, for $n = \max(n1, n2)$ [11]. Moreover, the construction of the $\chi(v)$ vectors will take $O(n\sqrt{n}L)$ time, implying that the overall complexity of the algorithm is $\max(O(n^2\sqrt{n\log\log n}), O(n^2\sqrt{n}L)$.

The approximation has to do with the use of a scaling parameter to find the maximum cardinality, minimum weight matching [11]; this parameter determines a tradeoff between accuracy and the number of iterations untill convergence. The matching matrix M in (4) can be constructed using the mapping set $\Phi(G,H)$. The algorithm is particularly well-suited to the task of matching two shock trees since it can find the best correspondence in the presence of occlusion and/or noise in the tree.

6 Experiments

6.1 Indexing Experiments

We test our indexing algorithm on a database of 60 object silhouettes, some representative examples of which are shown in Fig. 7. In the first experiment, we select 20 shapes from the database, compute their shock trees, compute the topological signature vectors for each of their nodes, and populate the resulting vectors in a model database. Each element, in turn, will be removed from the database and used as a query tree for the remaining database of 19 model trees. For each of the 20 trials, the 19 object candidates will be ranked in decreasing order of accumulator contents. To evaluate the quality of the indexing results, we will compute the distance between the query tree and each of the candidates, using the matcher developed in [31], and note which model tree is the closest match. If indexing is to be effective, the closest matching model should be among the best (highest-weight) candidates returned by the indexing strategy.

In the second and third experiments, we apply the same procedure to databases of size 40 and 60 model trees, respectively, in order to evaluate the scaling properties of our indexing algorithm. Thus, in the second experiment, we have 40 indexing trials, while in the third experiment, we have 60 indexing trials. Finding the position of the closest model shape among the sorted candidates for any given query shape requires that we first compute the 60 × 60 distance matrix.

Fig. 7. Samples from a Database of 60 Object Silhouettes

The results of the first experiment are shown in Fig. 8(a), where the horizontal axis indicates the rank of the target object (or closest matching object) in the sorted candidates, and the vertical axis represents the number of times that rank is achieved. For this experiment, the average rank is 1.6, which implies that on average, 8.4% of the sorted candidates need to be verified before the closest matching model is found. The results of the second and third experiments are shown in Fig. 8(b) and (c), respectively. The results are very encouraging and show that as database size increases, the indexing algorithm continues to prune over 90% of the database (avg. rank of 7.9% in expt. 2, 8.8% in expt. 3).

In a final experiment, we generate some occluded scenes from shapes in our database. In Table 1, we show three examples of occluded query images (left column) and the top ten (sorted left to right) model candidates from a database of 40 model shapes. Twice, the rank of the target is 4th, while once it is 3rd, indicating that for these three examples, at most 10% of the model indexing candidates need to be verified. We are currently conducting a more comprehensive set of occlusion experiments.

It should be noted that the indexing mechanism reflects primarily the topological structure of the query. Thus, in row 1 of Table 1, for example, the topological structure of the query (brush occluding the hammer) is more similar to the pliers-like objects (two of the top three candidates) than to the hammer itself. Topological similarity of shock trees is a necessary but not sufficient condition

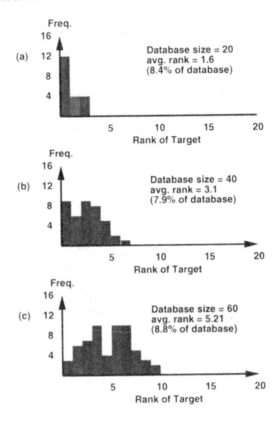

Fig. 8. Indexing Results for a Database of 60 Images. In each case, the horizontal axis indicates the rank of the target object (or closest matching object) in the sorted candidates, and the vertical axis represents the number of times that rank is achieved. (See text for discussion.)

for shape similarity, as it ignores the geometries of the object's parts (nodes in its shock tree). Therefore, the fact that objects with different shape can rank high in the candidate list is not surprising.

6.2 Matching of Saliency Maps

To evaluate our representation and matching framework, we apply it to a database of model object views generated by Murase and Nayar at Columbia University. Views of each of the 20 objects are taken from a fixed elevation every 5 degrees (72 views per object) for a total of 1440 model views. The top row of images in Fig. 9 shows three adjacent model views for one of the objects (piggy bank) plus one model view for each of two other objects (bulb socket and cup). The second row shows the computed saliency maps for each of the five images, while the third row shows the corresponding saliency map graphs. The time to compute the saliency map averaged 156 seconds/image for the five images on

Table 1. Indexing using an occluded query shape. For each row, the query is shown to the left, while the top ten candidate models (from a database of 40 models) are shown on the right, in decreasing order of weight.

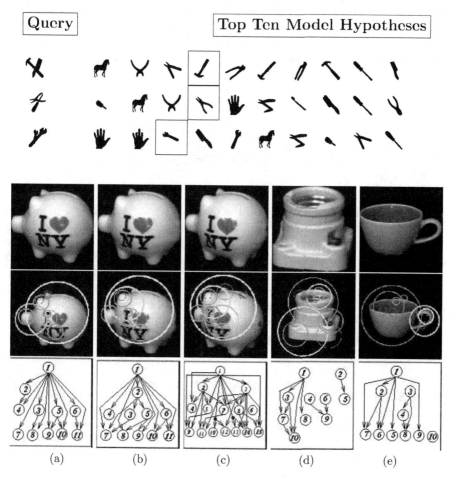

Fig. 9. A sample of views from the database: top row represents original images, second row represents saliency maps, while third row represents saliency map graphs.

a Sun Sparc 20, but can be reduced to real-time on a system with hardware support for convolution, e.g., a Datacube MV200. The average time to compute the distance between two SMG's is 50 ms using SMGBM, and 1.1 second using SMGAT (an average of 15 nodes per SMG).

To illustrate the matching of an unoccluded image to the database, we compare the middle piggy bank image (Fig. 9(b)) to the remaining images in the database. Table 2 shows the distance of the test image to the other images in Fig. 9; the two other piggy bank images (Fig.s 9 (a) and (c)) were the closest matching views in the entire database. Table 2 also illustrates the difference between the two matching algorithms. SMGBM is a weaker matching algorithm, searching for a topological match between two SMG's. SMGAT, on the other hand, is more restrictive, searching for a geometrical match between the two SMG's. For similar views, the two algorithms are comparable; however, as two views diverge in appearance, their similarity as computed by SMGAT diverges more rapidly than their SMGBM similarity. In a third experiment, we compare

Table 2. Distance of Fig. 9(b) to other images in Fig. 9

Algorithm	9(a)	9(c)	9(d)	9(e)
SMGBM	9.57	10.06	14.58	23.25
SMGAT	8.91	12.27	46.30	43.83

every image to every other image in the database, resulting in over 1 million trials. There are three possible outcomes: 1) the image removed from the database is closest to one of its neighboring views of the correct object; 2) the image removed from the database is closest to a view belonging to the correct object but not a neighboring view; and 3) the image removed from the database is closest to a view belonging to a different object. The results are shown in Table 3. As we would expect, the SMGAT algorithm, due to its stronger matching criterion, outperforms the SMGBM algorithm. If we include as a correct match any image belonging to the same object, both algorithms (SMGBM and SMGAT) perform extremely well, yielding success rates of 97.4% and 99.5%, respectively. To illus-

Table 3. An exhaustive test of the two matching algorithms. For each image in the database, the image is removed from the database and compared, using both algorithms, to every remaining image in the database. The closest matching image can be either one of its two neighboring views, a different view belonging to the correct object, or a view belonging to a different object.

Algorithm	% Hit	% Miss right object	% Miss wrong object
SMGBM	89.0	8.4	2.6
SMGAT	96.6	2.9	0.5

trate the matching of an occluded image to the database, we compare an image containing the piggy bank occluded by the bulb socket, as shown in Fig. 10. Table 4 shows the distance of the test image to the other images in Fig. 9. The closest matching view is the middle view of the piggy back which is, in fact, the view embedded in the occluded scene. In a labeling task, the subgraph matching the closest model view would be removed from the graph and the procedure applied to the remaining subgraph. After removing the matching subgraph, we

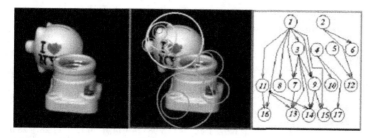

Fig. 10. Occluded Object Matching: (a) original image; (b) saliency map; and (c) saliency map graph

match the remaining scene subgraph to the entire database, as shown in Table 5. In this case, the closest view is the correct view (Figure 9(d)) of the socket.

Table 4. Distance of Fig. 10(a) to other images in Fig. 9. The correct piggy bank view (Fig. 9(b)) is the closest matching view.

Algorithm	9(a)	9(b)	9(c)	9(d)	9(e)
SMGBM	9.56	3.47	8.39	12.26	14.72
SMGAT	24.77	9.29	21.19	30.17	33.61

Table 5. Distance of Fig. 10(a) (after removing from its SMG the subgraph corresponding to the matched piggy back image) to other images in Fig. 9.

Algorithm	9(a)	9(b)	9(c)	9(d)	9(e)
SMGBM	12.42	14.71	14.24	4.53	9.83
SMGAT	18.91	20.85	17.08	7.19	15.44

6.3   Matching of Shock Trees

To evaluate our matcher's ability to compare objects based on their prototypical or coarse shape, we begin with a database of 24 objects belonging to 9 classes. To select a given class prototype, we select that object whose total distance to the other members of its class is minimum.[8] We then compute the similarity between each remaining object in the database and each of the class prototypes, with the results shown in Table 6. For each row in the table, a box has been placed around the most similar shape. We note that for the 15 test shapes drawn

---

[8] For each of the three classes having only two members, the class prototype was chosen at random.

Table 6. Similarity between database shapes and class prototypes. In each row, a box is drawn around the most similar shape (see the text for a discussion).

Instance	Distance to Class Prototype								
	⌇	⌇	⌇	Y	⌇	⌇	✋	🐎	⬤
⌇	⟦0.02⟧	2.17	4.48	3.55	2.96	0.21	4.58	14.33	10.01
⌇	2.39	⟦0.10⟧	5.97	15.90	3.98	0.14	26.12	17.28	28.94
⌇	10.89	4.72	⟦2.08⟧	12.24	3.12	2.15	19.73	10.11	12.64
⌇	7.15	6.42	⟦1.19⟧	1.35	5.10	3.38	10.58	11.11	11.11
⌇	4.08	7.72	2.98	⟦1.49⟧	4.26	4.14	26.60	13.54	14.21
⌇	14.77	6.72	5.69	⟦0.36⟧	2.30	5.90	10.58	16.25	19.10
⌇	7.86	8.90	5.94	⟦0.74⟧	1.59	1.10	10.81	10.39	16.08
⌇	2.66	4.23	3.23	6.47	⟦0.62⟧	1.48	11.73	15.38	15.15
⌇	3.18	5.31	1.25	4.64	⟦0.60⟧	1.30	14.18	17.22	9.08
⌇	4.55	0.76	1.32	2.86	1.49	⟦0.11⟧	21.38	15.35	13.04
✋	6.77	19.46	22.11	13.27	8.21	29.50	⟦0.15⟧	5.12	5.03
✋	8.73	23.14	31.45	24.41	10.16	31.08	⟦0.18⟧	8.45	7.05
🐎	12.46	19.0	27.40	14.58	24.26	17.10	8.85	⟦7.49⟧	16.93
⬤	13.86	23.07	12.81	11.24	17.48	23.23	6.02	6.92	⟦3.06⟧
⬤	15.73	21.28	14.10	12.46	19.56	19.21	9.53	7.12	⟦5.06⟧

from 9 classes, all but one are most similar to their class prototype, with the class prototype coming in a close second in that case.

Three very powerful features of our system are worth highlighting. First, the method is truly generic: the matching scores impose a partial ordering in each row, which reflects the qualitative similarity between structurally similar shapes. An increase in structural complexity is reflected in a higher cost for the best match, e.g., in the bottom two rows of Fig. 6. Second, the procedure is designed to handle noise or occlusion, manifest as missing or additional vertices in the shock graph. Third, the depth-first search through subtrees is extremely efficient.

7   Selected Related Work

Multi-scale image descriptions have been used by other researchers to locate a particular target object in the image. For example, Rao et al. use correlation to compare a multi-scale saliency map of the target object with a multi-scale saliency map of the image in order to fixate on the object [23]. Although these approaches are effective in finding a target in the image, they, like any template-based approach, do not scale to large object databases. Their bottom-up descriptions of the image are not only global, offering little means for segmenting an image into objects or parts, but offer little invariance to occlusion, object deformation, and other transformations.

Wiskott *et al.* [33] use Gabor wavelet jets to extract salient image features. Wavelet jets represent an image patch (containing a feature of interest) with a set of wavelets across the frequency spectrum. Each collection of wavelet responses represents a node in a grid-like planar graph covering overlapping regions of the image. Image matching reduces to a form of elastic graph matching, in which the similarity between the corresponding Gabor jets of nodes is maximized. Correspondence is proximity-based, with nodes in one graph searching for (spatially) nearby nodes in another graph. Effective matching therefore requires that the graphs be coarsely aligned in scale and image rotation.

Another related approach is due to Crowley et al. [3,2,4]. From a Laplacian pyramid computed on an image, peaks and ridges at each scale are detected as local maxima. The peaks are then linked together to form a tree structure, from which a set of *peaks paths* are extracted, corresponding to the branches of the tree. During matching, correspondence between low-resolution peak paths in the model and the image are used to solve for the pose of the model with respect to the image. Given this initial pose, a greedy matching algorithm descends down the tree, pairing higher-resolution peak paths from the image and the model. Using a log likelihood similarity measure on peak paths, the best corresponding paths through the two trees is found. The similarity of the image and model trees is based on a very weak approximation of the trees' topology and geometry, restricted, in fact, to a single path through the tree.

Graph matching is a very popular topic in the computer vision community. Although space prohibits us from providing a comprehensive review, we will mention some particularly relevant related work. A graduated assignment algorithm has been proposed for subgraph isomorphism, weighted graph matching, and attributed relational graph matching [13]. The method was applied to matching non-hierarchical point features and performs well in the presence of noise and occlusion. Cross and Hancock propose a two step matching algorithm for locating point correspondences and estimating geometric transformation parameters between 2-D images. Point correspondence is achieved via maximum a posteriori graph-matching, while expectation maximization (EM) is used to recover the maximum likelihood transformation parameters. The novel idea of using graph-based models to provide structural constraints on parameter estimation is an important contribution their work. This, combined with the EM algorithm, allows their system to impose an explicit deformational model on the feature points.

The matching of shock trees has been addressed by a number of other groups. In recent work, Pelillo et al. [21] introduced a matching algorithm which extends the detection of maximum cliques in association graphs to hierarchically organized tree structures. They use the concept of connectivity to derive an association graph, and prove that attributed tree matching is equivalent to finding a maximum clique in the association graph. They applied their algorithm to articulated and deformed shapes represented as shock trees. In a related paper, Tirthapura et al. [32] present an alternative use of shock graphs for shape matching. Their approach relies on graph transformations based on the edit distance between two graphs, defined as the "least action" path consisting of a sequence of elementary edit transformations taking one graph to another. The first approach can handle occlusion, but does not accommodate spurious noise in the graphs; the second approach handles spurious noise, but cannot effectively deal with occlusion. Both approaches focus solely on graph (tree) structure, and would have to be modified to include the concept of node similarity.

8   Conclusions

In this paper, we have reviewed three different algorithms for object recognition, each based on solving a bipartite matching formulation of a particular problem. The formulation is both very general and very powerful. We have shown edge weights that encode difference in region saliency, Euclidean distance in the image, and a function of topological and geometric distance. We have also seen different ways in which hierarchical ordering of nodes in a graph/tree can be enforced. In the case of saliency map graph matching, parent/child relationships are used to bias edge weights at lower levels of the matching, while in the case of shock tree matching, a depth-first procedure is used to ensure hierarchical consistency. It should be noted that the method by which we enforce hierarchical ordering in the matching of saliency map graphs is not applicable to the matching of shock graphs (DAGs or trees), since the method assumes that corresponding nodes in the hierarchy are at comparable scales. In a shock graph, a leaf child of the root may be as small in scale as a leaf further down the tree. However, we are exploring the application of our shock tree matching and indexing methods to multi-scale DAG representations.

Finally, we have shown how matching complexity can be managed in a coarse-to-fine framework. In the case of saliency map graph matching, solutions to the bipartite matching problem at a coarser level are used to constrain solutions at a finer level, while in the case of shock tree matching, large corresponding subtree roots (found through a solution to the bipartite matching problem) are used to establish correspondence between their descendents. Furthermore, in the case of shock tree matching, our eigen-characterization of a tree's topological structure allows us to efficiently compare subtree structures in the presence of noise and occlusion.

References

[1] Alizadeh, F.: Interior point methods in semidefinite programming with applications to combinatorial optimization. SIAM J. Optim. **5(1)** (1995) 13–51.

[2] Crowley, J., Parker, A.: A representation for shape based on peaks and ridges in the difference of low-pass transform. IEEE Transactions on Pattern Analysis and Machine Intelligence **6(2)** (1984) 156–169.

[3] Crowley, J. L.: A Multiresolution Representation for Shape. In Rosenfeld, editor. Multiresolution Image Processing and Analysis (1984) 169–189. Springer Verlag. Berlin.

[4] Crowley, J. L., Sanderson, A. C.: Multiple Resolution Representation and Probabilistic Matching of 2–D Gray–Scale Shape. IEEE Transactions on Pattern Analysis and Machine Intelligence **9(1)** (1987) 113–121.

[5] Cvetković, D.:Graphs and their spectra. University of Beograd (1971) 354–356.

[6] Cvetković, D., Rowlinson, P., Simić, S.: Eigenspaces of Graphs. Cambridge University Press. Cambridge, United Kingdom (1997).

[7] Mjolsness, G. G. E., Anandan., P.: Optimization in model matching and perceptual organization. Neural Computation **1** (1989) 218–229.

[8] Edmonds. E.: Paths, trees, and flowers. Canadian Journal of Mathematics **17** (1965) 449–467.

[9] Edmonds, J.,Matula. D.: An algorithm for subtree identification. SIAM Rev. **10** (1968) 273–274.

[10] Flynn, P., Jain, A.: 3D object recognition using invariant feature indexing of interpretation tables. CVGIP: Image Understanding **55(2)** (1992) 119–129.

[11] Gabow, H., Goemans, M., Williamson, D.: An efficient approximate algorithm for survivable network design problems. Proc. of the Third MPS Conference on Integer Programming and Combinatorial Optimization (1993) 57–74.

[12] Garey, M.,Johnson, D.: Computer and Intractability: A Guide to the Theory of NP-Completeness. Freeman. San Francisco (1979).

[13] Gold, S., Rangarajan, A.: A graduated assignment algorithm for graph matching. IEEE Transactions on Pattern Analysis and Machine Intelligence **18(4)** (1996) 377–388.

[14] Hopcroft, J., Karp, R.: An $n^{\frac{5}{2}}$ algorithm for maximum matchings in bipartite graphs. SIAM J. Comput. **2** (1973) 225–231.

[15] Kimia, B. B., Tannenbaum, A., Zucker, S. W.: Shape, shocks, and deformations I: The components of two-dimensional shape and the reaction-diffusion space. International Journal of Computer Vision **15** (1995) 189–224.

[16] Kobler, J.: The graph isomorphism problem: its structural complexity. Birkhauser. Boston (1993).

[17] Lamdan, Y., Schwartz, J., Wolfson., H.: Affine invariant model-based object recognition. IEEE Transactions on Robotics and Automation **6(5)** (1990) 578–589.

[18] Neumaier, A.: Second Largest Eigenvalue of a Tree. Linear Algebra and its Applications **46** (1982) 9–25.

[19] Nosal, E.: Eigenvalues of Graphs. University of Calgary (1970).

[20] Overton, M. L., Womersley, R. S.: Optimality conditions and duality theory for minimizing sums of the largest eigenvalues of symmetric matrices. Math. Programming **62(2)** (1993)321–357.

[21] Pelillo, M., Siddiqi, K., Zucker, S.W.: Matching hierarchical structures using association graphs. Fifth European Conference on Computer Vision **2** (1998) 3–16.

[22] Preparata, F., Shamos, M.: Computational Geometry. Springer-Verlag. New York, NY (1985).

[23] Rao, R. P. N, Zelinsky, G. J., Hayhoe, M. M., Ballard, D. H.: Modeling Saccadic Targeting in Visual Search. In Touretzky, Mozer, and Hasselmo, editors. Advances in Neural Information Processing Systems **8** 830–836. MIT Press. Cambridge, MA (1996).

[24] Reyner, S. W.: An analysis of a good algorithm for the subtree problem. SIAM J. Comput. **6** (1997) 730–732.

[25] Shokoufandeh, A., Marsic, I.,Dickinson, S.: Saleincy regions as a basis for object recognition. In Third International Workshop on Visual Form. Capri, Italy (1997).

[26] Shokoufandeh, A., Marsic, I., Dickinson., S.: View-based object matching. In Proceedings, IEEE International Conference on Computer Vision. Bombay (1998) 588–595.

[27] Shokoufandeh, A., Marsic, I., Dickinson., S.: View-based object recognition using saliency maps. Technical Report DCS-TR-339, Department of Computer Science, Rutgers University. New Brunswick, NJ 08903 (1998).

[28] Shokoufandeh, A., Marsic, I., Dickinson., S.: View-based object recognition using saliency maps. Image and Vision Computing **27** (1999) 445-460.

[29] Siddiqi, K., Kimia, B. B.: A shock grammar for recognition. Technical Report LEMS 143. LEMS, Brown University (1995).

[30] Siddiqi, K., Shokoufandeh, A., Dickinson, S., Zucker, S. W.: Shock graphs and shape matching. In Proceedings, IEEE International Conference on Computer Vision. Bombay (1998) pages 222–229.

[31] Siddiqi, K., Shokoufandeh, A., Dickinson, S., Zucker, S.W.: Shock graphs and shape matching. International Journal of Computer Vision **30** (1999) 1-22.

[32] Tirthapura, S., Sharvit, D., Klein, P., Kimia, B.B.: Indexing Based on Edit-Distance Matching of Shape Graphs. SPIE Proceedings on Multimedia Storage and Archiving Systems III (1998) 25–36.

[33] Wiskott, L., Fellous, J. M., Krüger, N., von der Malsburg, C.: Face Recognition by elastic bunch graph matching. IEEE Transactions on Pattern Analysis and Machine Intelligence **19(7)** (1997) 775–779.

An Introduction to Low-Density Parity-Check Codes

Amin Shokrollahi

Digital Fountain, Inc.
39141 Civic Center Drive
Fremont, CA 94538
USA
amin@digitalfountain.com

Abstract. In this paper we will survey some of the most recent results on low-density parity-check codes. Our emphasis will be primarily on the asymptotic theory of these codes. For the most part, we will introduce the main concepts for the easier case of the erasure channel. We will also give an application of these methods to reliable content delivery.

1   Introduction

The theory of Low-Density Parity-Check (LDPC) codes has attracted a lot of attention lately. This is mainly due to two reasons: (1) They are equipped with very efficient encoding and decoding algorithms, and (2) These algorithms are amenable to a theoretical analysis which has led to codes that operate at rates extremely close to theoretical bounds established by Shannon.

The situation with these codes is quite different from many other known classes of codes, e.g., algebraic codes. Traditionally, a code is shown to have good performance using non-constructive arguments. Once this is established, one tries to design efficient algorithms that match the performance predicted by the theory. LDPC codes are quite the opposite, as they are already equipped with efficient algorithms. The task here is to find those codes in the class that perform very well using these algorithms. This path has proved to be quite fruitful.

LDPC codes were discovered in the early 1960's by Gallager in his PhD-thesis [6]. They are constructed using sparse bipartite graphs. The construction is discussed in Section 3.

With the exception of excellent work by a few researchers like Zyablov, Pinsker, and Margulis [12,22,23] in Soviet Union and Tanner [21] in the US, LDPC codes were forgotten until the mid 1990's. The discovery of Turbo codes [3] in the coding community, and the search for efficiently encodable and decodable codes in the Theoretical Computer Science community led to a revival of LDPC codes [11,19,20]. However, it was not until the work by Luby et al. [10] on the erasure channel that researchers started to look at LDPC codes as codes that could achieve capacity using low-complexity decoders. [10] introduced a simple linear time erasure decoding algorithm over the erasure channel and showed that

G.B. Khosrovshahi et al. (Eds.): Theoretical Aspects of Computer Science, LNCS 2292, pp. 175–197, 2002.
© Springer-Verlag Berlin Heidelberg 2002

the only parameter that matters for the algorithm to perform well is the distribution of nodes of various degrees in the underlying graph. Moreover, using that analysis, the paper exhibited explicit degree distributions such that, in the limit, the corresponding codes achieve the capacity of the erasure channel. The analysis was further simplified in [8], and applied to simple decoding algorithms of Gallager for the binary symmetric channel in [9]. This analysis was taken up by Richardson and Urbanke in [15] and generalized to a very large class of channels. That work was extended in [14], which, following the example of [10], also exhibited degree distributions extremely close to channel capacity. The results were further refined in [4] to find codes with efficient decoding algorithms that correct fractions of errors closer to the Shannon capacity than any other known codes with efficient decoding.

The purpose of the present manuscript is to give an introduction into the theory of LDPC codes. Many of the deeper concepts have been described through the simpler case of codes over the erasure channel. Some care has been taken to keep the level of the presentation elementary while providing some crucial details.

The structure of this paper is as follows. In Section 2 we introduce the basic concepts: linear codes and communication channels. Section 3 introduces the construction of LDPC codes. The important class of message passage decoding algorithms is exemplified for the case of the erasure channel in Section 4. The algorithm is analyzed in Section 5 in the asymptotic case. Perhaps one of the most striking results about LDPC codes is that they achieve the capacity of the erasure channel using this simple algorithm. This is discussed in Section 6. After this introduction into LDPC codes and decoding algorithms, we are ready to discuss the general case of message passage decoding algorithms for LDPC codes in Section 7. Section 8 describes efficient encoding algorithms for LDPC codes. It may seem odd that encoding is discussed at the end of the paper. The reason for this is that efficient encoding algorithms use the erasure decoding algorithm. In Section 9 we show how LDPC codes over the erasure channel could be used to efficiently deliver content to a large number of receivers on the Internet.

As was indicated above, this note is meant as a brief introduction to the field of LDPC codes. We do not make any claims as to the completeness of these notes. In fact, many interesting results have been omitted for the sake of a short presentation. Nevertheless, we hope that these notes spark the reader's interest to learn more about this important class of error-correcting codes.

2   Codes and Channels

A k-dimensional linear code of block-length n over a field $\mathbb{F}$ is a k-dimensional subspace of the vector space $\mathbb{F}^n$. The elements of the code are called *codewords*. Often, we will refer to vectors in $\mathbb{F}^n$ as words and to their coordinates as their *symbols*.

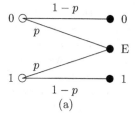
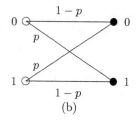

Fig. 1. (a) The binary erasure channel, and (b) the binary symmetric channel

Any $k \times n$-matrix whose rows from a basis for the code is called a *generator matrix*. Such a matrix G is in *systematic form* if it is of the form $(I_k \mid B)$, where I_k is the $k \times k$-identity matrix. The word *systematic* is explained by the *encoding map*: A message $m = (m_1, \dots, m_k) \in \mathbb{F}^k$ is encoded to $(c_1, \dots, c_n) := m \cdot G$. If G is in systematic form, the first k coordinates of the codeword are identical to the message symbols. We call the quantity k/n the *rate* of the code. This is the amount per symbols of genuine information carried by a codeword.

To check whether a vector $c = (c_1, \dots, c_n)$ belongs to the code, we can use any *parity-check matrix* of the code: An $(n - k) \times n$-matrix whose rows from the dual space of the code with respect to the standard scalar product is called a parity-check matrix for the code. Then c belongs to the code if and only if $H \cdot c = 0$.

In the following we will concentrate on binary codes, i.e., codes over the field $\mathbb{F}_2$ with two elements. This is the most important case for the protection of information in the physical layer of communication media (e.g., wireless or optical channels). We will however also talk about codes for the network layer where the individual symbols are network packets which consist of many bits. We will see, however, that most of what we will talk about also holds for that case (with modifications).

In practice the exact nature of errors imposed on codewords during a communication is not completely known and is approximated by different channel models. For the purposes of this paper, the following definition of a communication channel is adequate. A channel is a finite labeled directed bipartite graph between a set A called the *code-alphabet* and a set B called the *output-alphabet* such that the labels are nonnegative real numbers and satisfy the following property: for any element $a \in A$, the sum of the labels of edges emanating from a is 1. Semantically, the graph describes a communication channel in which elements from A are transmitted and those of B are received. The label of an edge from a to b is the (conditional) probability of obtaining b given that a was transmitted. Examples of Channels are given in Figure 1.

How reliably can we transmit over such a channel? This question was answered by Shannon in 1948. To state it, we need first to introduce the notion of maximum-likelihood decoding. A maximum-likelihood codeword associated to a word $y \in \mathbb{F}^n$ is a codeword c which differs from y in the smallest number of coordinate positions. Maximum-likelihood decoding is the task of computing a maximum-likelihood codeword associated to a given word. This task is NP-hard in general [2] but we will postpone efficiency issues until later. It is not clear that the maximum-likelihood decoding does in fact produce the codeword that had been transmitted. However, it is the receiver's best guess based on the available information. To assess how much coding can improve reliability of communication, it is thus imperative to know how large the probability of error of this algorithm is. This fundamental question was answered by the famous *Channel Coding Theorem*: for any channel there is a parameter C, called its *capacity*, such that for any rate $R < C$ there is a sequence of codes of rate R of increasing block-length such that the error probability of the maximum likelihood decoding for these codes approaches 0 as the block-length goes to infinity. In fact, the error probability is smaller than e^{-cn} for some constant c depending on the channel and the rate, where n is the block-length.

Computing the capacity of a channel is not easy in general, but not very hard for simple channels like the BEC and the BSC. For instance, the capacity of the BEC with erasure probability p is $1 - p$, while that of the BSC with the cross-over probability p is $1 - H(p)$, where $H(p) = -p \log p - (1 - p) \log(1 - p)$ is the binary entropy function.[1]

Shannon's paper answered a number of questions but at the same time also generated many new ones. The first question is how to explicitly construct the codes promised by the Channel Coding Theorem. The second more serious question is that of efficient decoding of such codes, as maximum-likelihood decoding is a very hard task in general.

Low-density parity-check (LDPC) codes, described in the next section, are very well suited to answer these questions.

3   Construction of LDPC Codes

In the following we will assume that the code-alphabet A is the binary field $\mathbb{F}_2$. Let G be a bipartite graph between n nodes on the right called *variable nodes* (or message nodes) and r nodes on the right called *check nodes*. The graph gives rise to a code in (at least) two different ways, see Figure 2: in Gallager's original version [6] the coordinates of a codeword are indexed by the variable nodes $1, \dots, n$ of G. A vector $(x_1, \dots, x_n)$ is a valid codeword if and only if for each check node the XOR[2] of the values of its adjacent variable nodes is zero. Since each check node imposes one linear condition on the x_i, the rate of the code is at least $(n - r)/n$.

In the second version, the variable nodes are indexed by the original message. The check nodes contain the redundant information: the value of each such node is equal to the sum (over $\mathbb{F}_2$) of the values of its adjacent variable nodes. The block-length of this code is $n + r$, and its rate is $n/(n + r)$.

---

[1] In this paper log denotes logarithm to the basis 2.

[2] In this paper, XOR denotes the sum modulo 2, and is sometimes denoted by $\oplus$.

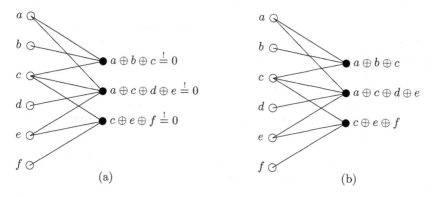

Fig. 2. The two versions of LDPC codes: (a) Original version, and (b) dual version

These two versions look quite similar, but differ fundamentally from a computational point of view. The encoding time of the second version is proportional to the number of edges in the graph G. Hence, if G is sparse, the encoding time is linear in the number of variable symbols (for constant rate). On the other hand, it is not clear how to encode the first version without solving systems of linear equations. We will see later in Section 8 that for many types of Gallager codes the encoding can be essentially performed in linear time. (This is a result due to Richardson and Urbanke [16].)

In this paper, will only consider Gallager's original construction of codes from graphs. Following his notation, we will call these codes Low-Density Parity-Check (LDPC) codes, thereby indicating the sparseness of the underlying graph.

4   Decoding on the Erasure Channel

The aim of this section is to develop a decoding algorithm for LDPC codes on the (binary) erasure channel. If the erasure probability of the channel is denoted by p, then with probability p a symbol is erased during transmission. The model of an erasure channel was introduced by Elias [5]. As was mentioned earlier, the capacity of this channel is $1 - p$ [5]. By Shannon's coding theorem, This means that to achieve reliable transmission, the rate of the code used has to be less than $1 - p$. It should be intuitively clear that if the rate is higher than $1 - p$, then we cannot expect to recover anything: after a p-fraction of the information is lost, only a $1-p$-fraction remains and if the rate is less than $1-p$, the amount of information in the remaining part is strictly less than the amount of original information so no recovery is possible.

Our aim now is to design, for any given p, codes of rates arbitrarily close to $1-p$ that can recover a fraction of p or less erasures and do so in time linear in the number of variable symbols. The notion of running time has to be made precise: Here, the running time of our algorithm is the number of arithmetic operations it performs over the ground field (usually $\mathbb{F}_2$). In particular, we do not count access to memory (even though this can be quite burdensome in applications). We only mention that counting the latter results in increasing the running time of our algorithms by an additional log-factor if we use the model of a RAM with logarithmic cost measure.

Following Elias, we will first discuss the general case of a linear code. The minimal Hamming weight of a nonzero element in a linear code is called the *minimum distance* of the code, usually denoted by d. The following result is easy to prove [5].

Proposition 1. *A linear code of minimum distance d is capable of correcting any pattern of $d - 1$ or less erasures with $O((n - k)^3)$ operations over the base field.*

PROOF.   We start from a parity-check matrix H of the code. It has $n - k$ rows and n columns. We claim that if the code has minimum distance d, then any $d - 1$ columns of H are linearly independent. Otherwise, there is a nontrivial linear dependency among $d - 1$ columns of H, i.e., the right kernel of H (which is by definition the original code), has a nontrivial element of weight $d - 1$ or less, which is a contradiction to d being the minimum distance of the code.

To be able to correct $d - 1$ or fewer erasures, it is sufficient to describe an algorithm which, on input $n - d + 1$ or more symbols of a codeword, can reconstruct the codeword. After permuting the coordinate places, we may assume that the first $n - d + 1$ positions are known, while the last $d - 1$ are unknown. Let H_1 and H_2 be the matrices consisting of the first $n - d + 1$ and the last $d - 1$ columns of H, respectively. Let $y_1, \dots, y_{n-d+1}$ denote the received symbols of the codeword and $x_1, \dots, x_{d-1}$ denote the erased symbols. Then we have $H_1(y_1, \dots, y_{n-d+1})^\top = H_2(x_1, \dots, x_{d-1})^\top$, which gives a system of $n - k$ equations in $d - 1$ unknowns. Since the columns of H_2 are independent, this system has at most one solution which consists of the erased coordinates. $\square$

To achieve capacity on the erasure channel with erasure probability p using the above algorithm, we need $d - 1$ to be roughly equal to pn, and the rate to be roughly equal to $(1 - p)n$. It can be shown that for codes over the alphabet $\mathbb{F}_2$ this relationship cannot hold for large n and positive p [7, Chap. 5]. A closer look reveals that in our decoding algorithm we insisted to be able to correct *any* pattern of erasures. This is rather restrictive (and certainly not in the spirit of the Coding Theorem). There is much more to be gained if we allow for a small probability of decoding error.

Elias [5] showed that random linear codes achieve capacity of the erasure channel with high probability. In a nutshell, the argument is as follows: let p be the error probability of the channel, and choose a random linear code of block-length n and dimension $(1-p)n - \sqrt{n}$, say. (We ignore diaphontine constraints.) Then, a $(pn + \sqrt{n}) \times pn$-submatrix of a parity-check matrix of the code has full rank pn, with high probability. Now we can apply the above erasure decoding algorithm to recover from a p-fraction of erasures by solving a system of $pn + \sqrt{n}$ equations in pn unknowns. The running time of this algorithm is clearly of the order $O(n^3)$ (if we do not use fast but impractical matrix multiplication algorithms). This is very slow for applications in which n is very large (in the order of 100,000's; in Section 9 we will describe an application in which this range for the block-length is not unusual). To find more efficient decoding algorithms, we thus have to look elsewhere.

It turns out that LDPC codes can be decoded very efficiently on the binary erasure channel. The decoding algorithm can be described as follows.

Algorithm 1 *The decoder maintains a register for each of the variable and check nodes. All of these registers are initially set to zero.*

1. *[Direct recovery step.] XOR the value of each received variable node v to the values of its adjacent check nodes, and delete v and all edges emanating from it from the graph.*
2. *[Substitution recovery step.] If there exists a check node c of degree one, do:*
   a) Copy the value of c to its adjacent variable node v.
   b) Delete v and all edges emanating from the graph.
3. *If c does not exist and the graph is not empty, output* Decoder Failure *and stop.*
4. *If c does not exit because the graph is empty, output* Decoder success *and stop.*
5. *Go back to Step 2.*

Note that the decoding time is proportional to the number of edges in the graph. If the graph is sparse, i.e., if the number edges is linear in the number of nodes, then the decoder is linear time (at least on a RAM with unit cost measure).

The hope is that there is always enough supply of degree one check nodes so that the decoder finishes successfully. Whether or not this is the case depends on the original fraction of erasures, and on the graph. Surprisingly, however, the only important parameter of the underlying graph is the distribution of nodes of various degrees. This analysis is the topic of the next section.

5   Analyzing Erasure Correction

One key component to the analysis of the erasure correction is the introduction of probability generating functions. Let X be a random variable over the set of nonnegative integers. For every integer n, the probability $\Pr[X = n] =: f_n$ is thus a real number less than 1, and $\sum_{i=0}^{\infty} f_n = 1$. The *generating function* of X is the power-series

$$f_X(s) := \sum_{i=0}^{\infty} f_n s^n.$$

For the analysis of the erasure decoder, we will mostly be dealing with generating functions of 0/1-random variables. In this case, the generating function of the random variable X can be written as

$$f_X(s) = f_0 + (1 - f_0)s.$$

A moment's thought reveals that if X and Y are independent 0/1-random variables with generating functions $f(s) = f_0 + (1 - f_0)s$ and $g(s) = g_0 + (1 - g_0)s$, respectively, then the logical OR $X \wedge Y$ and the logical AND $X \wedge Y$ of X and Y have the generating functions

$$f_{X \vee Y}(s) = f_0 g_0 + (1 - f_0 g_0)s, \qquad f_{X \wedge Y}(s) = (1 - f_1 g_1) + f_1 g_1 s, \qquad (1)$$

respectively.

Often we are also interested in the random variables $S_M = X_1 \vee X_2 \vee \cdots \vee X_M$ and $P_M = X_1 \wedge X_2 \wedge \cdots \wedge X_M$, where the X_i are mutually independent identically distributed 0/1-random variables, with the common distribution $f(s)$, and the number M is a random variable independent of the X_i. In this case, we have the following simple result for the generating functions of S_M and P_M. We will assume that the empty OR and the empty AND are 0 and 1, respectively.

Proposition 2. *Suppose that* $X_1, \ldots, X_M$ *are mutually independent 0/1-random variables with the common distribution* $f(s) = f_0 + (1 - f_0)s$, *where* M *is a random variable with generating function* $g(s) = \sum_{n \geq 0} g_n s^n$ *which is independent of the* X_i. *Further, let* $F(s)$ *and* $G(s)$ *denote the generating functions of the random variables* $S_M = X_1 \vee X_2 \vee \cdots \vee X_M$ *and* $P_M = X_1 \wedge X_2 \wedge \cdots \wedge X_M$, *respectively. Then we have*

$$F(s) = g(f_0) + (1 - g(f_0))s$$
$$G(s) = (1 - g(1 - f_0)) + g(1 - f_0)s.$$

PROOF.   We have

$$\Pr[S_M = 0] = \Pr[M = 0] + \sum_{n \geq 1} \Pr[M = n]\Pr[X_1 \vee \cdots \vee X_n = 0],$$

and

$$\Pr[P_M = 1] = \Pr[M = 0] + \sum_{n \geq 1} \Pr[M = n]\Pr[X_1 \wedge \cdots \wedge X_n = 0],$$

where we have used the convention that the empty OR is 0 and the empty AND is 1. By (1) the probability $\Pr[X_1 \vee \cdots \vee X_n = 0]$ equals f_0^n, while $\Pr[X_1 \wedge \cdots \wedge X_n = 1] = (1 - f_0)^n$. Hence, $\Pr[S_M = 0] = g(f_0)$, while $\Pr[P_M = 1] = g(1 - f_0)$. The result follows. $\square$

Now we are ready for heuristic analysis of the erasure decoder. Let us first re-describe the decoder.

At each round, the decoder sends 0/1-messages from variable nodes to check nodes and back. Furthermore, to each variable node there is associated a channel information, itself a 0/1-value, which states whether the node was an erasure (value 0) or not (value 1). The message from a variable node v to a check node c is obtained from the logical OR of the channel information and all messages from all the check nodes incident to v other than c. Semantically, this means that the message sent along the edge (v, c) is a 1 if and only if the node v was recovered. At a check node, the message sent from node c to variable node v is the logical AND of the messages from all the variable nodes incident to c other than v. Semantically, this message is a 1 if and only if the check node c can recover the variable node v. This happens if and only if all the variable nodes incident to c other than v have already been recovered. In round 0 of the algorithm the variable nodes send their associated channel information to all their incident check nodes.

The heuristic analysis of this algorithm is accomplished by tracking the generating functions of the messages passed along edges at each round of the algorithm. For this, we assume that the underlying graph is a graph whose edges have generating functions $\lambda(s)$ and $\rho(s)$. What we mean by this is the following: for each i, let λ_i denote the probability that an edge is connected to a variable node of degree i, and let ρ_i denote the probability that an edge is connected to a check node of degree i. Then we define

$$\lambda(s) = \sum_{i \geq 1} \lambda_i s^{i-1}, \qquad \rho(s) = \sum_{i \geq 1} \rho_i s^{i-1}.$$

The reason for considering the exponent $i - 1$ rather than i will become clear in a moment.

Let $f^n(s) = p_n + (1 - p_n)s$ and $h^n(s) = q_n + (1 - q_n)s$ denote the generating functions of the the random variables $X_{v,c}^n$ and $Y_{c,v}^n$ describing the messages passed from variable node v to check node c and vice-versa at round n. Suppose that each variable node is erased with probability p_0. The messages passed from variable nodes to check nodes in round 0 of the algorithm all have the generating function $f^0(s)$. Let v be a variable node and c be a check node incident to v. The message $Y_{c,v}^n$ passed from c to v at round n is the logical AND of the messages passed to c from all the variable nodes incident to c other than v. These have the common generating function $f^n(s)$. *If we assume that the random variables of the incoming messages are independent*, then, by Proposition 2, the generating function of $Y_{c,v}^n$ is $q_n + (1 - q_n)s$ where $q_n = (1 - g(1 - p_n)) + g(1 - p_n)s$, and $g(s)$ is the generating function of the random variable describing the degree of c minus 1. The probability that the edge (c, v) is connected to a check node of degree i is ρ_i by assumption; in that case, one has to take the logical OR of $i - 1$ variables. Hence, $g(s) = \rho(s)$ and the generating function of $Y_{c,v}^n$ is $q_n + (1 - q_n)s$ with $q_n = 1 - \rho(1 - p_n)$.

As for the random variable $X_{v,c}^n$, this is the logical OR of the random variables $Y_{c',v}^{n-1}$ and the random variable X_0 describing the channel information, where c' runs over all the check nodes incident to v other than c. Again, under the assumption of the independence of the random variables $Y_{c',v}^{n-1}$ for all c', and using the same reasoning as above, we see that $X_{v,c}^n$ has the generating function $p_n + (1 - p_n)s$ where

$$p_n = p_0 \lambda(q_{n-1}) = p_0 \lambda(1 - \rho(1 - p_{n-1})).$$

For successful decoding we need p_n to be less than p_{n-1} in each round. (In fact, we need $p_n \leq (1 - \epsilon)p_{n-1}$ for all n a fixed ϵ.) From this we obtain the following criterion for successful decoding:

$$\forall\, x \in (0, p_0]:\qquad p_0\lambda(1 - \rho(1 - x)) < x. \tag{2}$$

In other words, given $\lambda(s)$ and $\rho(s)$, the code can tolerate no more than a p_0-fraction of erasures, where p_0 is the supremum of all the real numbers satisfying (2). In fact, if the analysis were correct, then we could deduce that the expected maximum fraction of losses the code can tolerate is p_0. However, the independence assumption underlying the analysis is not valid in general, which casts serious doubts on the correctness of our criterion. In the following we will sketch how to turn this analysis into an exact one, at least in an asymptotic setting. For now, we consider an example.

Example 1. Suppose that the underlying graph is biregular of bidegree (3,6). That is, all the variable nodes are of degree 3, while all the check nodes are of degree 6. A random graph of this type gives rise to a code of rate at least 1/2. In this case $\lambda(x) = x^2$ and $\rho(x) = x^5$. The maximum tolerable loss fraction would be the maximum p_0 such that

$$p_0(1 - (1 - x)^5)^2 < x$$

for $x \in (0, p_0]$. This maximum p_0 that satisfies this inequality is roughly equal to 0.429 (see [1] for an exact computation of p_0). Despite the non-rigorous analysis above, this value is very close to the one obtained through extensive simulations for large block-lengths!

The above example is quite typical. That is, given a degree sequence, the threshold value p_0 for that sequence is very close to the one predicted by the above analysis for large block-lengths. This can be explained rigorously, though we will only sketch the rigorous proof and point the reader to relevant literature [9,15]. The analysis relies in part on the fact that if the neighborhoods around variable nodes are trees, then the independence assumption is valid. For a random graph and fixed depth d, it is easily shown that, when the block-length is large, then neighborhoods of depth d of almost all variable nodes are tree-like (almost all = all but an inverse polynomial fraction). Hence, the independence assumption is valid for up to d rounds of the algorithm.

The independence assumption is only part of the argument. The other part consists of showing that the performance of any instantiation of the decoding process is close to the expected performance given by (2). This is done by setting up an appropriate martingale and using Azuma's inequality to prove the concentration of sums of random variables with limited dependencies around their expectation. For a detailed discussion of these techniques the reader is referred to [9,15]. Putting things together, we see that for a fixed depth d, there is an n_0 such that a for a random graph with $n \geq n_0$ variable nodes, a d-fold iteration of the erasure decoding algorithm reduces the fraction of errors below $(1 - \epsilon)^d p_0$. By choosing d large enough, we can push this fraction below any constant fraction. To show that the process finishes from this point on, we can use expansion properties of random graphs [10].

The asymptotic analysis sketched above seems not to be the last word on this subject. To get the neighborhoods of variable nodes up to even moderate depths tree-like one would need codes of extremely large block-lengths. However, the successful experiments reported above are for much smaller block-lengths. This seems to suggest that the dependencies of the random variables involved is possibly not too serious. An exact formulation of this statement and a proof thereof are certainly among the foremost open problems in the field.

6   Achieving Capacity

The condition (2) is very handy if one wants to analyze the performance of random graphs with a *given* degree distribution. However, it does not give a clue on how to design good degree distributions λ and ρ. Our aim is to construct sequences that asymptotically achieve the capacity of the erasure channel. In other words, we want p_0 in (2) to be arbitrarily close to $1 - R$, where R is the rate of the code. In the following, we call a sequence $(\lambda^\ell, \rho^\ell)_{\ell \geq 0}$ of degree distributions *capacity-achieving of rate R* if

- The corresponding graphs give rise to codes of rate at least R,
- For all $\epsilon > 0$ there exists an ℓ_0 such that for all $\ell \geq \ell_0$ we have

$$(1 - R)(1 - \epsilon)\lambda(1 - \rho(1 - x)) < x \quad \text{for} \quad x \in (0, (1 - R)(1 - \epsilon)).$$

Before discussing these sequences, it makes sense to relate the rate of the code to the degree distributions $(\lambda(x), \rho(x))$. Note that the average left and right degree of the graph are

$$\frac{1}{\sum_i \frac{\lambda_i}{i}} = \frac{1}{\int_0^1 \lambda(x)\mathrm{d}x}, \quad \text{and} \quad \frac{1}{\sum_i \frac{\rho_i}{i}} = \frac{1}{\int_0^1 \rho(x)\mathrm{d}x},$$

respectively. As a result, the rate of the code is at least $1 - \int_0^1 \rho(x)\mathrm{d}x / \int_0^1 \lambda(x)\mathrm{d}x$. It is a nice exercise to deduce from the equation (2) alone that p_0 is always less than or equal to $1 - R$, i.e., less than or equal to $\int_0^1 \rho(x)\mathrm{d}x / \int_0^1 \lambda(x)\mathrm{d}x$.

The first capacity-achieving sequence for the erasure channel was discovered in [10]: fix a parameter D and let $\lambda_D(x) := \frac{1}{H(D)} \sum_{i=1}^D x^i/i$, where $H(D)$ is the harmonic sum $\sum_{i=1}^D 1/i$. Note that

$$\int_0^1 \lambda_D(x)\mathrm{d}x = \frac{1}{H(D)}\left(1 - \frac{1}{D+1}\right).$$

Let $\rho_D(x) := e^{\mu(x-1)}$, where μ is the unique solution to the equation

$$\int_0^1 \rho_D(x)\mathrm{d}x = \frac{1}{\mu}(1 - e^{-\mu}) = (1 - R)\int_0^1 \lambda_D(x)\mathrm{d}x = \frac{1-R}{H(D)}\left(1 - \frac{1}{D+1}\right).$$

Then the sequence $(\lambda_D(x), \rho_D(x))_{D \geq 1}$ gives rise to codes of rate at least R. Further, we have

$$p_0 \lambda_D(1 - \rho_D(1 - x)) = p_0 \lambda_D(1 - e^{-\mu x})$$
$$\leq \frac{-p_0}{H(D)}\ln(e^{-\mu x})$$
$$= \frac{p_0 \mu x}{H(D)}.$$

Hence, successful decoding is possible if the fraction of erasures is no more than $H(D)/\mu$. Note that this quantity equals $(1-R)(1-1/(D+1))/(1-e^{-\mu})$, which is larger than $(1-R)(1-1/D)$. Hence, we have

$$(1-R)\left(1-\frac{1}{D}\right)\lambda_D\left(1-\rho_D(1-x)\right) < x \quad \text{for} \quad x \in \left(0, (1-R)(1-1/D)\right).$$

This shows that the sequence is indeed capacity-achieving. This sequence of degree distributions is called the *Tornado sequence* and the derived codes are called *Tornado codes*.

Another capacity-achieving sequence discovered in [17] is closely related to the power series expansion of $(1-x)^{1/D}$. For integers $a \geq 2$ and $n \geq 2$ let

$$\rho_a(x) := x^{a-1}, \qquad \lambda_{a,n}(x) := \frac{\displaystyle\sum_{k=1}^{n-1}\binom{\alpha}{k}(-1)^{k+1}x^k}{1 - n\binom{\alpha}{n}(-1)^{n+1}},$$

where $\alpha := 1/(a-1)$. Then it can be shown that with the right choice of the parameter n (dependent on a) this sequence is indeed capacity-achieving.

For the Heavy-Tail/Poisson sequence the average degree of a variable node was less than $H(D)$, and it could tolerate up to $(1-R)(1-1/D)$ fraction of erasures. Hence, to get close to within $1-\epsilon$ of the capacity $1-R$, we needed codes of average degree $O(\log(1/\epsilon))$. This is shown to be essentially optimal in [17]. In other words, to within $1-\epsilon$ of the channel capacity, we need graphs of average degree $\Omega(\log(1/\epsilon))$. The same relation also holds for the right-regular sequences. Hence, these codes are essentially optimal for the erasure decoder.

It would be interesting to obtain a concise description of all capacity-achieving sequences. However, this seems to be too much to ask for. In [13] the authors start a systematic study of such sequences. The first observation is that both for the Tornado and the right-regular sequences, we start with a function $f(x)$ represented by a Taylor series with non-negative coefficients on $[0,1]$ and satisfying the normalizations $f(0)=0$, $f(1)=1$, for which

$$\mathcal{T}f(x) := 1 - f^{-1}(1-x)$$

has again a converging Taylor series expansion with non-negative coefficients. Therefore, [13] concentrates on the set

$$\mathcal{P} := \left\{ f(x) = \sum_{1}^{\infty} f_k x^k, \ x \in [0,1] \ \middle| \ f_k \geq 0, f(1) = 1 \right\},$$

and the set $\mathcal{A}$ defined as the maximal subset of $\mathcal{P}$ invariant under the action of $\mathcal{T}$:

$$\mathcal{A} := \{ f \in \mathcal{P} \mid \mathcal{T}f \in \mathcal{P} \}.$$

[13] proves various properties of these function sets and derive new capacity-achieving sequences. One of the results of that paper is that the right-regular

sequence is capacity-achieving in a much stronger sense than the Tornado sequence. For more details the reader is referred to [13].

Another interesting property of the examples of capacity-achieving sequences given above is their "flatness:" for the examples above, the derivatives of the function $f_\ell(x) := p_0 \lambda^\ell(1-\rho^\ell(1-x)) - x$ converge to zero on any open subinterval of $(0, p_0)$. In [18] it was proved that this condition is necessary for *any* capacity-achieving sequence. More precisely, [18] shows that for any fixed k, the derivatives of $f_\ell(x)$ up to order k converge uniformly to 0 on any open subinterval of $(0, p_0)$. In particular, by continuity, the first derivative of f_ℓ converges to 0 at 0. In other words, we have

$$p_0 \lambda_2^\ell \frac{\mathrm{d}\rho^\ell(x)}{\mathrm{d}x}\bigg|_{x=1} \to 1$$

as $\ell \to \infty$. This seems to suggest that there is no capacity-achieving sequence such that almost all derived graphs do not have variable nodes of degree 2. The non-existence of variable nodes of degree 2 has important algorithmic consequences, as these are the last nodes to be decoded in the process. (This should be intuitively clear: variable nodes of degree 2 are connected to only two check nodes, so has lowest probability of being decoded. The argument can be made precise.)

7   Codes on Other Channels: Message Passage Decoding

The erasure decoder described in Algorithm 1 belongs to the class of *message passing decoders*. These are decoders obeying the following rules. At each round of the algorithm messages are passed from variable nodes to check nodes and back. The criterion is that the message sent from the variable node v to the check node c only depends on the messages sent from check nodes c' to v where c' runs over all the check nodes incident to v *other than* c. The erasure decoder is an instance of a message passing decoder.

The first examples of message passing decoders were given by Gallager [6] for the binary symmetric channel. A simple decoding algorithm in this case is as follows: In round zero of the algorithm the variable nodes send their received value to their incident check nodes. The message sent from check node c to variable node v is obtained by XOR'ing the messages received by c from all incident variable nodes other than v. The message sent from the variable node v to the check node c is the received value of node v unless all the messages received by v from all incident check nodes other than c agree and are equal to b, say; in this case, the message sent from v to c is equal to b.

The operation at the check nodes is clear: the messages sent to variable nodes are chosen such that the XOR of all the incoming messages is zero. On the variable side the messages sent are rather conservative: the variable node changes its value only if all the incident check nodes tell it to do so.

This algorithm can be analyzed along the same lines as the erasure decoding algorithm. We associate to the 0/1-random variable X the generating function $f_X(s) = q + ps$, where $p = \Pr[X = 0]$ and $q = 1 - p$. If X and Y are independent random variables with generating functions $f_X(s)$ and $f_Y(s)$, then the generating function of $Z = X \oplus Y$ is given by $f_X(s)f_Y(s) \bmod (s^2 - 1)$, where $\oplus$ is the XOR operation. This is easily checked: the probability that Z is zero is $\Pr[X = 0]\Pr[Y = 0] + \Pr[X = 1]\Pr[Y = 1]$. Similarly, $\Pr[Z = 1] = \Pr[X = 0]\Pr[Y = 1] + \Pr[Y = 0]\Pr[X = 1]$. It turns out that these two quantities are precisely the coefficients of the linear polynomial $f_X(s)f_Y(s)$.

Now suppose that $X_1, \ldots, X_m$ are independent random variables having the same generating function $f(s) = q + ps$. Then the generating function of $Z = X_1 \oplus \cdots \oplus X_m$ is given by $f(s)^m \bmod (x^2 - 1)$. This is most conveniently computed using the Hadamard-Transform: suppose that $H(f) = (p + q, q - p) = (1, 1 - 2p)$. Then $H(f^m) = (1, (1 - 2p)^m)$. Further, the inverse of H is given by $H^{-1}((x, y)) = \frac{1}{2}(x + y) + \frac{1}{2}(x - y)s$. So, the generating function of the XOR of $X_1, \ldots, X_m$ equals

$$\frac{1 + (1 - 2p)^m}{2} + \frac{1 - (1 - 2p)^m}{2}s.$$

To proceed with the analysis of the algorithm, suppose that the messages sent from variable nodes to check nodes at round i of the algorithm have the common generating function $q_i + p_i s$. Then the message sent from check nodes back to variable nodes at that round have the density

$$\frac{1 + \rho(1 - 2p_i)}{2} + \frac{1 - \rho(1 - 2p_i)}{2}s.$$

This follows in the same way as in the case of the erasure decoder. At the variable nodes we have the following probability distribution of the message passed from variable node v to check node c. Suppose that the incoming messages all have the common generating function $a_i + b_i s$, and that the initial noise has the generating function $q_i + p_i s$. The probability that the message passed from v to c is 0 is equal to the probability that all the incoming messages from incident nodes other than c are zero, plus the probability that the initial noise was zero (a_i) multiplied with the probability that not all the incoming messages agree. Denoting by Z the random variable describing the message sent from v to c, we have then

$$\Pr[Z = 1] = \lambda(b_i)(1 - p_0) + p_0(1 - \lambda(a_i)),$$

where p_0 is the probability that the node received the value 1. Altogether, we obtain the following identity:

$$p_{i+1} = (1 - p_0)\lambda\left(\frac{1 - \rho(1 - 2p_i)}{2}\right) + p_0\left(1 - \lambda\left(\frac{1 + \rho(1 - 2p_i)}{2}\right)\right).$$

Assuming that the transmitted word was the all-zero word, the condition for successful decoding becomes

$$(1 - p_0)\lambda\left(\frac{1 - \rho(1 - 2x)}{2}\right) + p_0\left(1 - \lambda\left(\frac{1 + \rho(1 - 2x)}{2}\right)\right) < x,$$

for $x \in (0, p_0]$. For example, for the $(3,6)$-graph, the largest p_0 satisfying this inequality is roughly 0.0394. (See [1] for the exact value of this threshold.) By Shannon's Theorem, the highest sustainable error probability for a rate $1/2$ code is roughly 11%. Hence, this code is not particularly interesting. A more appealing code is obtained by a $(4, 8)$-graph. In this case, the threshold p_0 can be computed to be equal to $\frac{1}{21}$. In general, for a $(d, 2d)$-graph with $d > 3$, the threshold turns out to be

$$\frac{1}{11(d - 3) + 4\binom{d-3}{2} + 10}.$$

In [1] the authors discuss the best degree distributions for various rates for the above decoding algorithm. In particular, they exhibit a degree sequence with threshold roughly equal to 0.0513.

The above heuristic analysis can be made rigorous in an asymptotic setting in exactly the same way as above.

The decoding algorithm described above belongs to the class of hard-decision decoding algorithms: the messages passed back and forth are bits. To achieve performance closer to capacity, one needs to use more refined information. The most powerful decoding algorithm for LDPC codes is the belief-propagation, also discovered by Gallager in [6]. The description of the algorithm is as follows.

We will use the standard map $0 \mapsto 1$, $1 \mapsto -1$. At each iteration, messages are passed along the edges of the graph from variable nodes to their incident check nodes and back. The messages are typically real valued but they can also take on the values $\pm\infty$, reflecting the situation where some bits are known with absolute certainty.

Generically, messages which are sent in the ℓ-th iteration will be denoted by $\mathrm{m}^{(\ell)}$. By $\mathrm{m}_{vc}^{(\ell)}$ we denote the message sent from the variable node v to its incident check node c, while by $\mathrm{m}_{cv}^{(\ell)}$ we denote the message passed from check node c to its incident variable node v. Each message represents a quantity $\ln(p^+/p^-)$, where $p^+ = p(\mathrm{x} = 1 \mid y)$, $p^- = p(\mathrm{x} = -1 \mid y)$, x is the random variable describing the codeword bit value associated to the variable node v, and y is the random variable describing all the information incorporated into this message. By Bayes rule we have

$$\mathrm{m} = \ln \frac{p(\mathrm{x} = 1 \mid y)}{p(\mathrm{x} = -1 \mid y)} = \ln \frac{p(y \mid \mathrm{x} = 1)}{p(y \mid \mathrm{x} = -1)},$$

since x is equally likely ± 1. The message m is the log-likelihood ratio of the random variable x (under the *independence assumption*).

As we will see shortly, to represent the updates performed by *check nodes* an alternative representation of the messages is appropriate. Let us define a map $\gamma : [\infty, +\infty] \to \mathbb{F}_{(2)} \times [0, +\infty]$ as follows. Given $x \in [\infty, +\infty]$, $x \neq 0$, let

$$\gamma(x) := (\gamma_1(x), \gamma_2(x)) := (\mathrm{sgn}\, x, -\ln \tanh |\frac{x}{2}|). \tag{3}$$

Let m_0 be the log-likelihood ratio of the codeword bit $\mathrm{x} = \pm 1$ associated to the variable node v conditioned only on the channel observation of this bit. The update equations for the messages under belief-propagation are then the following:

$$\mathrm{m}_{vc}^{(\ell)} = \begin{cases} \mathrm{m}_0, & \text{if } \ell = 0, \\ \mathrm{m}_0 + \sum_{c' \in C_v \setminus \{c\}} \mathrm{m}_{c'v}^{(\ell)}, & \text{if } \ell \geq 1, \end{cases} \tag{4}$$

$$m_{cv}^{(\ell)} = \gamma^{-1}\left(\sum_{v'\in V_c\setminus\{v\}} \gamma(m_{v'c}^{(\ell-1)})\right),\qquad (5)$$

where C_v is the set of check nodes incident to variable node v, and V_c is the set of variable nodes incident to check node c.

Now let f_ℓ denote the probability density function at the variable nodes at the ℓth round of the algorithm. f_0 is then the density function of the error which the variable bits are originally exposed to. It is also denoted by P_0. These density functions are defined on the set $\mathbb{R}\cup\{\pm\infty\}$. It turns out that they satisfy a *symmetry condition* [14] $f(-x) = f(x)e^{-x}$. As a result, the value of any of these density functions is determined from the set of its values on the set $\mathbb{R}_{\geq0}\cup\{\infty\}$. The restriction of a function f to this set is denoted by $f^{\geq0}$. (The technical difficulty of defining a function at ∞ could be solved by using distributions instead of functions, but we will not further discuss it here.)

For a function f defined on $\mathbb{R}_{\geq0}\cup\{\infty\}$ we define a *hyperbolic change of measure* γ via

$$\gamma(f)(x) := f(\ln\coth x/2)\operatorname{csch}(x).$$

If f is a function satisfying the symmetry condition, then $\gamma(f^{\geq0})$ defines a function on $\mathbb{R}_{\geq0}\cup\{\infty\}$ which can be uniquely extended to a function F on $\mathbb{R}\cup\{\pm\infty\}$. The transformation mapping f to F is denoted by Γ. It is a bijective mapping from the set of density functions on $\mathbb{R}\cup\{\pm\infty\}$ satisfying the symmetry condition into itself. Let f_ℓ denote the density of the common density function of the messages passed from variable nodes to check nodes at round ℓ of the algorithm. f_0 then denotes the density of the original error, and is also denoted by P_0. Suppose that the graph has a degree distribution given by $\lambda(x)$ and $\rho(x)$. Then we have the following:

$$f_\ell = P_0 \otimes \lambda(\Gamma^{-1}(\rho(\Gamma(f_{\ell-1})))),\quad \ell\geq1.\qquad (6)$$

Here, $\otimes$ denotes the convolution, and for a function f, $\lambda(f)$ denotes the function $\sum_i \lambda_i f^{\otimes(i-1)}$. In the case of the erasure channel, the corresponding density functions are two-point mass functions, with a mass p_ℓ at zero and a mass $(1-p_\ell)$ at infinity. In this case, the iteration translates to [14]

$$p_\ell = \delta\lambda(1-\rho(1-p_{\ell-1})),$$

where δ is the original fraction of erasures. This is exactly the same as in (2).

The iteration (6) is proved in a similar way as the iteration for the erasure channel: one assumes an independence assumption first, and then shows that for large graphs the assumption is valid for a fixed round of iterations. Details can be found in [15].

Using Equation (6) it is possible to design degree sequences such that the corresponding codes can tolerate a fraction of errors using the belief-propagation algorithm, which is very close to the Shannon bounds. For instance, for the binary symmetric channel, the Shannon bound asserts that a code of rate 1/2 cannot tolerate a fraction of errors more than 0.110028. Using the degree sequence

$$\lambda(x) := 0.157581x + 0.164953x^2 + 0.0224291x^3 + 0.045541x^4 + 0.0114545x^5 +$$
$$0.0999096x^6 + 0.0160667x^7 + 0.00258277x^8 + 0.00454797x^9 +$$
$$0.000928767x^{10} + 0.0188361x^{11} + 0.0648277x^{12} + 0.0206867x^{13} +$$
$$0.000780516x^{14} + 0.0383603x^{15} + 0.0419398x^{16} + 0.0023117x^{19} +$$
$$0.00184157x^{20} + 0.0114194x^{22} + 0.0116636x^{28} + 0.0850183x^{39} +$$
$$0.01048x^{40} + 0.0169308x^{55} + 0.0255644x^{56} + 0.0364086x^{70} +$$
$$0.0869359x^{74}$$

$$\rho(x) := 0.25x^9 + 0.75x^{10},$$

we obtain a code that can asymptotically recover from a fraction 0.106 of errors using the belief-propagation algorithm. This and many other examples can be found in [14].

8 Encoding

Recall the two definitions of LDPC codes from Section 3. In Version (b) of the construction encoding is rather trivial: each right node is obtained by XOR'ing the values of its adjacent left nodes. In this case, however, the simple erasure decoding algorithm discussed above does not necessarily work: if we wanted to use the algorithm in this setting, we would need to assume that the losses did not occur among the right nodes. There is no reason why this should be the case. To remedy this situation, one can protect the redundant right nodes by another graph. Again, in this case one would need to protect the rightmost nodes, which leads to the construction of yet another graph, etc. The cascade of graphs obtained this way can eventually be closed by a powerful binary code. Details can be found in [10].

One problem with this construction is that if the fraction of errors in even one of the graphs comprising the cascade is larger than what the code can tolerate, then the decoding algorithm fails. Even though the cascading idea is very elegant and appealing in theory, it quickly faces its limitations in practice. Although there are some suggestions to remedy the situation [8], a practical solution is not available at this point.

Surprisingly, there are practical solutions for Gallager's construction of LDPC codes even though this may not seem at first sight. To understand the solution, we need to look at a matrix representation of LDPC codes. Much of the presentation of this section has been taken from [16].

The parity-check matrix of an LDPC code given by a bipartite graph with n left nodes and r right nodes is the $r \times n$-matrix over $\mathbb{F}_2$ whose columns are indexed by the left nodes and whose rows are indexed by the right nodes. There is a 1 at position (i, j) if and only if the jth left node is connected to the ith right node. This matrix is said to be *the* parity-check matrix of the LDPC code. In the following, we will call this matrix H, as for most of our discussions it is unimportant which kind of graph we are using.

The erasure decoding algorithm described previously can be described in terms of the matrix H in the following way:

Proposition 3. *Let C be an LDPC code of block-length n and co-dimension r, and let H be its parity-check matrix. Suppose that the erasure decoding algorithm 1 can recover from erasures in positions $i_1, i_2, \ldots, i_t$. Then, after a permutation of rows and columns, the submatrix of H corresponding to the columns $i_1, i_2, \ldots, i_t$ becomes lower triangular. Conversely, if the submatrix of H consisting of columns $i_1, i_2, \ldots, i_t$ is lower triangular up to a permutation of rows and columns, then the erasure decoding algorithm can recover erasures in positions $i_1, i_2, \ldots, i_t$.*

PROOF. Let G denote the graph corresponding to H. The rows of H correspond to the r right nodes of G while the columns correspond to the n left nodes. W.l.o.g., let us assume that the erasures are he t first positions. The subgraph induced by these nodes is obtained from H by deleting columns $t+1, t+2, \ldots, n$. Let us denote this matrix by M_1. By assumption, there is now a right node of degree 1, i.e., M_1 has a row of degree 1. Let us call the left node corrected by this right node v_1. After a permutation of rows and columns, we can assume that position $(1, 1)$ of M_1 is 1, and that position $(1, i)$ of M_1 is 0 for $i \geq 2$. Deleting the first row of M_1, we obtain the matrix M_2 which describes the graph in which v_1 and all edges incident to it are deleted. Again, by assumption, there is another right node of degree 1 in the graph new graph, which corrects, say, v_2. By a permutation of rows and columns we can assume that the $(1, 1)$-entry of M_2 is 1 and that the entries $(1, i)$ of M_2 are zero for $i \geq 2$. Continuing this way, we see that the original matrix M_1 is lower triangular. $\square$

Perhaps the simplest possible case for fast encoding is the following.

Lemma 1. *With the same assumptions as above, suppose that H contains an $r \times r$-submatrix which is lower triangular up to a permutation of rows and columns. Then the code corresponding to H can be encoded in linear time.*

PROOF. Without loss of generality assume that the $r \times r$-submatrix is the one given by columns $1, 2, \ldots, r$. Choosing positions $r+1, r+2, \ldots, n$ freely, we can recover the positions $1, 2, \ldots, r$ using the erasure decoding algorithm, since this is guaranteed to finish successfully by Proposition 3. The erasure decoding algorithm runs in linear time which implies that the full encoding algorithm does so as well. $\square$

In general one cannot expect the conditions of the lemma to be satisfied. However, one can hope that H contains an $m \times m$-submatrix which is lower triangular up to a permutation of rows, where m is close to r. In that case, one has the following.

Proposition 4. *With the same conditions as above, suppose that H contains an $m \times r$-submatrix which is lower triangular up to a permutation of rows and columns. Then the code corresponding to H can be encoded in time $O(n + (r - m)^2)$ after a preprocessing time of $O(n + (n + r - 2m)(r - m)^2 + m(n - m)(r - m))$, where n and r are the number of columns and rows of H, respectively.*

PROOF.   Without loss of generality, let us assume that the $m \times r$-submatrix M consists of the first m columns and that the rows have already been permuted so that M is lower triangular. Then H has the following form:

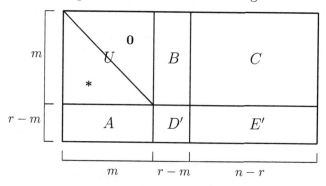

The matrix U is lower triangular, and the matrices A, B, C, D', and E' are sparse. For the encoding process we can assume that A is zero. This is seen by computing the Schur-complement of H with respect to U: Multiplying H from the left with the invertible matrix

$$\begin{pmatrix} I & 0 \\ -AU^{-1} & I \end{pmatrix},$$

where the I's are appropriately dimensioned identity matrices, then H becomes $\tilde{H}$ given as

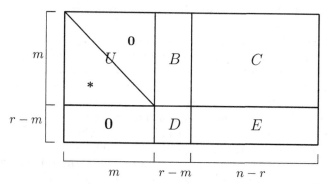

where $D = D' - AU^{-1}B$ and $E = E' - AU^{-1}C$. This multiplication does not change the right kernel of H, i.e., the code.

Let us assume for the moment that the matrix D is invertible. Then we compute in a preprocessing step the matrix D^{-1}. This takes $O((r-m)^3)$ time. In general, this matrix is not sparse anymore.

The encoding algorithm takes as input a vector $(x_1, x_2, \ldots, x_k)$, $k = n - r$, and produces an encoded vector $c = (c_1, c_2, \ldots, c_n)$ via the following algorithm, which successively determines the subvectors $d = (c_1, \ldots, c_m)$, $e = (c_{m+1}, \ldots, c_r)$, and $f = (c_{r+1}, \ldots, c_n)$ of c. We assume that we have computed the matrix D and that it is invertible.

1. Set $f := x$, i.e., $c_{r+1} := x_1, c_{r+2} := x_2, \ldots, c_n := x_k$.
2. Compute $e := D^{-1} \cdot (E \cdot f)$ by first determining $E \cdot f$ and then multiplying D^{-1} with it.
3. Compute $y := B \cdot e + C \cdot f$ and solve $U \cdot d = y$ for d.

To see that this algorithm works, note that the encoding process is equivalent to solving the equations

$$D \cdot e + E \cdot f = 0,$$
$$U \cdot d + B \cdot e + C \cdot f = 0.$$

Let us now discuss its running time.

Step 2. The computation of $E \cdot f$ is accomplished by realizing that this is equal to $(E' - AU^{-1}C)f$. The vector $C \cdot f$ is computed with $O(n-r)$ operations. Since U is sparse and lower triangular, U^{-1} times this vector can be computed in time $O(m)$. The vector obtained this way can be multiplied with A with $O(m)$ operations, as A is sparse. $E' \cdot f$ can be computed with $O(n-r)$ operations, as E' is sparse. Hence, $E \cdot f$ can be computed with $O(n)$ operations. Multiplication of D^{-1} with this vector uses $O((r-m)^2)$ operations, as D^{-1} is dense in general. So, this step uses $O(n + (r-m)^2)$ operations.

Step 3 uses $O(n-r)$ operations for the computation of y and $O(m)$ operations to solve for d, since U is sparse.

Altogether, we see that if D is invertible, then the running time of the algorithm is $O((r-m)^2 + n)$ The preprocessing of this algorithm consists of computing the matrix D. It is easy to see that this is accomplished with $O(m(r-m)^2)$ operations.

What if D is not invertible? In this case, we compute $\tilde{D} := (D' \mid E') - AU^{-1}(B \mid C)$ and run Gaussian elimination on this matrix appended with the $(r-m) \times (r-m)$-identity matrix to obtain a submatrix that is invertible. The computation of $\tilde{D}$ costs $O(m(n-m)(r-m))$. Running Gaussian elimination on this matrix costs $O((n+r-2m)(r-m)^2)$ operations. This step also reveals whether or not H has full rank, in which case we simply reduce r.

Note that in a practical setting a full Gaussian elimination on the compound matrix $\tilde{D}$ is not necessary; typically, appending $r - m$ columns of E to D will reveal an invertible submatrix. So, in practice the running time of the preprocessing step is likely to be a lot less than $(r-m)^3$. □

An immediate application of this proposition is that if the erasure decoding algorithm can correct a large fraction of errors for a code, then that code can be encoded fast!

Corollary 1. *Let C be an LDPC code of rate R and block-length n for which the erasure decoding algorithm can correct a $(1-R)(1-\epsilon)$-fraction of erasures. Then the code can be encoded in time $O(n + (\epsilon \cdot n)^2)$ time.*

PROOF.   Follows from the previous proposition and Proposition 3. □

Example 2.   1. Consider a random $(3,6)$-graph. We have seen that the erasure decoding algorithm can asymptotically correct a fraction δ of erasures, where $\delta \sim 0.429$. Hence, using the encoding algorithm above, the running time of the encoder is proportional to $n + 0.0049n^2$. In fact, the only nonlinear part of the running time is multiplication with a square matrix of size $0.071 \cdot n$ which results in an overall running time of $0.0049n^2$.

2. Consider the Tornado codes. Fix the design parameter D. In this case, the erasure decoding algorithm can asymptotically recover from a fraction $(1 - R)(1 - 1/D)$ of errors. It turns out that the running time of the encoding algorithm is roughly $(n/D)^2$.

One can apply the same type of techniques to the transpose $H^\top$ of H rather than to H. With that, it is possible to further lower the gap $r - m$ of the encoding process. For instance, for the $(3,6)$-graph, the encoding requires in addition to processes with linear running time, a multiplication with a square matrix of size $0.027n$. For details, we refer the reader to [16].

With the above method it is not possible to prove that the encoding time is linear, except if one can show that the erasure decoding algorithm can recover from a fraction $(1 - R)(1 - 1/sqrt(n))$ of erasures. The methods that we have discussed in this paper do not allow the proof of such an assertion. In [16] the authors improve upon the decoding algorithm described in this section and show that for a large class of LDPC codes decoding is indeed possible in linear time. More precisely, they show that if an LDPC code has a degree distribution given by $\lambda(x)$ and $\rho(x)$, and if the distribution satisfies the conditions

- $\rho(1 - \lambda(1 - x)) \leq x$ on $(0,1)$, and
- $\lambda'(0)\rho'(1) > 1$,

and if the minimum degree of check node is larger than 2, then the code can be encoded in linear time, with high probability. Here, the probability is with respect to the random choice of the graph. Somewhat surprisingly, all the optimized degree sequences from [14] satisfy the above properties, and so can be encoded in linear time with high probability.

9   An Example of the Erasure Channel: The Internet

The definition of the erasure channel may seem slightly out of touch with realities in the world of communication. This was certainly the case until the proliferation of Internet applications and the introduction of computer networks as a communication channel. The latter provides the prime example of the erasure channel.

In a computer network data is transferred in the form of atomic entities called packets. Packets are routed through the network from the transmitter to the receiver. During this transmission, packets can be lost or corrupted. Each packet is equipped with a header which, among other things, contains information on the entity the packet came from (the specific file it belongs to) and its position within that entity. Upon reception, the receiver puts the packets into their right order and hence can identify those that are lost. Another source of error on the

Internet is the corruption of packets. Each packet is equipped with a standard checksum which is designed to detect corrupted packets. Once a packet is known to be corrupted, it can be regarded as lost. Hence, one needs only to concentrate on lost packets.

Standard protocols like the TCP Protocol solve the problem of lost packets by initiating retransmission by the server. In a scenario in which the server is transmitting the same file to a huge number of receivers, this solution becomes inadequate. This is because the server needs to keep track of losses of all the individual clients. To solve the loss problem in this case, one needs to scale the number of servers linearly with the number of receivers. Financially, this becomes a losing proposition.

A completely different approach can be taken in the following way. The server encodes the file using an LDPC code especially designed for the erasure channel. An example could be the Tornado codes. Then the server broadcasts the encoded file in a cyclic manner into the network.

The client joins the broadcast and collects a number of packets which is slightly larger than the size of the original file. Once this is achieved, the erasure decoder starts and decodes the file.

In this scenario, there is no need for a feedback channel between the user and the server. The user waits until it has received enough distinct packets, no matter which ones.

The key to the realization of this solution is the existence of codes described in this paper. It is rather interesting that a theoretical solution like this leads to the design of a massively scalable content distribution architecture.

References

[1] L. Bazzi, T. Richardson, and R. Urbanke. Exact thresholds and optimal codes for the binary symmetric channel and Gallager's decoding algorithm A. *IEEE Trans. Inform. Theory*, 2001. Submitted.

[2] E.R. Berlekamp, R.J. McEliece, and H.C.A. van Tilborg. On the inherent intractability of certain coding problems. *IEEE Trans. Inform. Theory*, 24:384–386, 1978.

[3] C. Berrou, A. Glavieux, and P. Thitimajshima. Near Shannon limit error-correcting coding and decoding. In *Proceedings of ICC'93*, pages 1064–1070, Geneva, Switzerland, May 1993.

[4] S-Y. Chung, D. Forney, T. Richardson, and R. Urbanke. On the design of low-density parity-check codes within 0.0045 dB of the Shannon limit. *IEEE Commu. Letters*, 2001. To appear.

[5] P. Elias. Coding for two noisy channels. In *Information Theory, Third London Symposium*, pages 61–76, 1955.

[6] R. G. Gallager. *Low Density Parity-Check Codes*. MIT Press, Cambridge, MA, 1963.

[7] J.H. van Lint. *Introduction to Coding Theory*, volume 86 of *Graduate Texts in Mathematics*. Springer Verlag, third edition, 1998.

[8] M. Luby, M. Mitzenmacher, and A. Shokrollahi. Analysis of random processes via and-or tree evaluation. In *Proceedings of the 9th Annual ACM-SIAM Symposium on Discrete Algorithms*, pages 364–373, 1998.

[9] M. Luby, M. Mitzenmacher, A. Shokrollahi, and D. Spielman. Analysis of low density codes and improved designs using irregular graphs. In *Proceedings of the 30th Annual ACM Symposium on Theory of Computing*, pages 249–258, 1998.

[10] M. Luby, M. Mitzenmacher, A. Shokrollahi, D. Spielman, and V. Stemann. Practical loss-resilient codes. In *Proceedings of the 29th annual ACM Symposium on Theory of Computing*, pages 150–159, 1997.

[11] D.J.C. MacKay. Good error-correcting codes based on very sparse matrices. *IEEE Trans. Inform. Theory*, 45:399–431, 1999.

[12] G. A. Margulis. Explicit constructions of graphs without short cycles and low density codes. *Combinatorica*, 2:71–78, 1982.

[13] P. Oswald and A. Shokrollahi. Capacity-achieving sequences for the erasure channel. 2000. Manuscript.

[14] T. Richardson, A. Shokrollahi, and R. Urbanke. Design of provably good low-density parity check codes. *IEEE Trans. Inform. Theory (submitted)*, 1999.

[15] T. Richardson and R. Urbanke. The capacity of low-density parity check codes under message-passing decoding. *IEEE Trans. Inform. Theory (submitted)*, 1998.

[16] T. Richardson and R. Urbanke. Efficient encoding of low-density parity-check codes. *IEEE Trans. Inform. Theory*, 2001. Submitted.

[17] A. Shokrollahi. New sequences of linear time erasure codes approaching the channel capacity. In M. Fossorier, H. Imai, S. Lin, and A. Poli, editors, *Proceedings of the 13th International Symposium on Applied Algebra, Algebraic Algorithms, and Error-Correcting Codes*, number 1719 in Lecture Notes in Computer Science, pages 65–76, 1999.

[18] A. Shokrollahi. *Capacity-achieving sequences*, volume 123 of *IMA Volumes in Mathematics and its Applications*, pages 153–166. 2000.

[19] M. Sipser and D. Spielman. Expander codes. *IEEE Trans. Inform. Theory*, 42:1710–1722, 1996.

[20] D. Spielman. Linear-time encodable and decodable error-correcting codes. *IEEE Trans. Inform. Theory*, 42:1723–1731, 1996.

[21] M. R. Tanner. A recursive approach to low complexity codes. *IEEE Trans. Inform. Theory*, 27:533–547, 1981.

[22] V. V. Zyablov. An estimate of the complexity of constructing binary linear cascade codes. *Probl. Inform. Transm.*, 7:3–10, 1971.

[23] V. V. Zyablov and M. S. Pinsker. Estimation of error-correction complexity of Gallager low-density codes. *Probl. Inform. Transm.*, 11:18–28, 1976.

Primal-Dual Schema Based Approximation Algorithms

Vijay V. Vazirani[*]

College of Computing
Georgia Institute of Technology
Atlanta, GA 30332–0280
USA
vazirani@cc.gatech.edu

Abstract. The primal–dual schema, a general algorithm design method, has yielded approximation algorithms for a diverse collection of **NP**-hard problems. Perhaps its most remarkable feature is that despite its generality, it has yielded algorithms with good approximation guarantees as well as good running times. This survey provides some insights into how this comes about as well as a discussion on future work.

1   Introduction

A large fraction of the theory of approximation algorithms, as we know it today, is built around linear programming. Let us first provide reasons for this. When designing an approximation algorithm for an **NP**-hard **NP**-optimization problem, one is immediately faced with the following dilemma. In order to establish the approximation guarantee, the cost of the solution produced by the algorithm needs to be compared with the cost of an optimal solution. However, for such problems, not only is it **NP**-hard to find an optimal solution, but it is also **NP**-hard to compute the cost of an optimal solution. Typically, the way to get around this dilemma is to find good, polynomial time computable lower bound on the cost of the optimal solution (assuming we have a minimization problem at hand). Interestingly enough, the lower bounding method provides a key step in the design of the algorithm itself. Many combinatorial optimization problems can be stated as integer programs. Once this is done, the linear relaxation of this program provides a natural way of lower bounding the cost of the optimal solution. Approximation algorithms for numerous problems follow this approach.

There are two basic techniques for obtaining approximation algorithms using linear programming. The first, and more obvious, method is to solve the linear program and then convert the fractional solution obtained into a n integral solution, trying to ensure that in the process the cost does not increase much. The approximation guarantee is established by comparing the cost of the integral and fractional solutions. This technique is called *rounding*.

---

[*] Research supported by NSF Grant CCR-9820896

G.B. Khosrovshahi et al. (Eds.): Theoretical Aspects of Computer Science, LNCS 2292, pp. 198–207, 2002.

The second, less obvious and perhaps more sophisticated, method is to use the dual of the LP-relaxation in the design of the algorithm. This technique is called the *primal–dual schema*. Let us call the LP-relaxation the primal program. Under this schema, an integral solution to the primal program and a feasible solution to the dual program are constructed iteratively. Notice that any feasible solution to the dual also provides a lower bound on OPT. The approximation guarantee is established by comparing the two solutions.

For many problems, both techniques have been successful in yielding algorithms having good guarantees, often essentially equal to the integrality gap of the relaxation being used. The main difference in performance between the two techniques lies in the running times of the algorithms produced. An LP-rounding algorithm needs to find an optimal solution to the linear programming relaxation, and therefore has a high running time, often leading to an unpractical algorithm. Perhaps such inefficiency is to be expected from so general a method. The following quote by Lovasz [15] about another general algorithm design schema, due to Grotschel, Lovasz and Schrijver [9,10], which yields exact polynomial time algorithms for numerous problems, is quite insightful in this respect. "The polynomial time algorithms which follow from these general considerations are very far from being practical. ... This ridiculous slowness is of course quite natural for a method which can be applied in such generality – and hence makes so little use of the specialities of particular problems."

It is quite surprising then, that despite its generality, the primal–dual schema yields algorithms with good running times. The reason is that the primal–dual schema provides only a broad outline of the algorithm. The details have to be designed by utilizing the special combinatorial structure of individual problems. In fact, for many problems, once the algorithm has been designed using the primal–dual schema, the scaffolding of linear programming can be completely dispensed with to get a purely combinatorial algorithm.

This brings us to another advantage of the primal–dual schema – this time not objectively quantifiable. A combinatorial algorithm is more malleable than an algorithm that requires an LP-solver. Once a basic problem is solved using the primal–dual schema, one can also solve variants and generalizations of the basic problem. From a practical standpoint, a combinatorial algorithm is more useful, since it is easier to adapt it to specific applications and fine tune its performance for specific types of inputs.

In this survey paper, we will first present a broad historical development of the ideas behind this schema. Then, we will give the basic mechanism it uses and illustrate it in the simple setting of the set cover problem. We will also show how the algorithm, once designed, can be stated in purely combinatorial terms. Finally, we will list some key open problems concerning this schema.

2   Historical Development of Ideas

Kuhn [14] gave the first primal–dual algorithm – for the weighted bipartite matching problem. However, he used the name "Hungarian Method" to describe his algorithm. Dantzig, Ford and Fulkerson [5] used this method for giving another means of solving linear programs, and called it the *primal–dual method*. Although the schema was not very successful for solving linear programs, it soon found widespread use in combinatorial optimization. Indeed, this schema yielded the most efficient algorithms for some of the cornerstone problems in **P**, including matching, network flow and shortest paths. These problems have the property that their LP-relaxations have integral optimal solutions. By the LP-duality theorem we know that optimal solutions to linear programs are characterized by fact that they satisfy all the complementary slackness conditions. In fact, the primal–dual schema for exact algorithms is driven by these conditions. Starting with initial feasible solutions to the primal and dual programs, it iteratively starts satisfying complementary slackness conditions. When they are all satisfied, both solutions must be optimal. During the iterations, the primal is always modified integrally, so that eventually we get an integral optimal solution.

Consider an LP-relaxation for an **NP**-hard problem. In general, the relaxation will not have an optimal solution that is integral. Does this rule out a complementary slackness condition driven approach? Interestingly enough, the answer is "No". It turns out that the algorithm can be driven by a suitable relaxation of these conditions! This is the most commonly used way of designing primal–dual based approximation algorithms – but not the only way.

In retrospect, the first use of the primal–dual schema in approximation algorithms was due to Bar-Yehuda and Even [2]. They gave a factor two algorithm for the weighted vertex cover problem. The set cover algorithm given in Section 4 is a simple generalization of their idea. It is interesting to note that they did not originally state their work as a primal–dual algorithm.

The works of Agrawal, Klein and Ravi [1] and Goemans and Williamson [7] revived the use of this schema in the setting of of approximation algorithms, giving a factor 2 approximation algorithm for the Steiner forest problem. They also introduced the powerful idea of growing duals in a synchronized manner. This is has turned out to be an important way in which primal–dual algorithms differ in the exact and approximate setting. The former work *on demand*, in the sense that they pick a condition that needs to be fixed and do so. The paper of Goemans and Williamson also established the wide applicability of this schema by giving algorithms for several related problems such as the prize collecting Steiner tree problem.

The mechanism of relaxing complementary slackness conditions was first formalized in Williamson, Goemans, Mihail, and Vazirani [18] in the context of solving the Steiner network problem, a generalization of the Steiner tree problem to higher connectivity requirements. All of the above stated algorithms work with a covering-packing pair of linear programs, i.e., all the coefficients occurring in the LP's are nonnegative. The extension to non-covering-packing LP's was given by Jain and Vazirani [12] in the context of deriving an algorithm for the metric uncapacitated facility location problem. This work also introduced the use of Lagrangian relaxation, a classic method from combinatorial optimization, into the primal–dual schema, by giving an algorithm for solving the metric k-median problem. The latter problem has a global constraint – that at most k facilities be opened. On the other hand, the primal–dual schema works by making local improvements. The use of Lagrangian relaxation enables replacing the global constraint by a local one, thereby essentially "reducing" the k-median problem to the facility location problem. This technique has also been used for "reducing" the k-MST problem, which has a global constraint, to the prize collecting Steiner tree problem [4]. The only primal–dual algorithm that does not operate by relaxing complementary slackness conditions is [17]; they relax the dual program instead. For further historical information, see the excellent survey by Goemans and Williamson [8]. For a detailed exposition of several primal–dual schema based approximation algorithms, see [Vaz01].

3   Overview of the Schema

Let us consider the following primal program, written in standard form.

$$\text{minimize} \quad \sum_{j=1}^{n} c_j x_j$$

$$\text{subject to} \quad \sum_{j=1}^{n} a_{ij} x_j \geq b_i, \quad i = 1, \ldots, m$$

$$x_j \geq 0, \qquad j = 1, \ldots, n$$

where a_{ij}, b_i, and c_j are specified in the input. The dual program is:

$$\text{maximize} \quad \sum_{i=1}^{m} b_i y_i$$

$$\text{subject to} \quad \sum_{i=1}^{m} a_{ij} y_i \leq c_j, \quad j = 1, \ldots, n$$

$$y_i \geq 0, \qquad i = 1, \ldots, m$$

Most known approximation algorithms using the primal–dual schema run by ensuring one set of conditions and suitably relaxing the other. In the following description we capture both situations by relaxing both conditions. Eventually, if primal conditions are ensured, we set $\alpha = 1$, and if dual conditions are ensured, we set $\beta = 1$.

Primal complementary slackness conditions
  Let $\alpha \geq 1$.
  For each $1 \leq j \leq n$: either $x_j = 0$ or $c_j/\alpha \leq \sum_{i=1}^{m} a_{ij} y_i \leq c_j$.
Dual complementary slackness conditions
  Let $\beta \geq 1$.
  For each $1 \leq i \leq m$: either $y_i = 0$ or $b_i \leq \sum_{j=1}^{n} a_{ij} x_j \leq \beta \cdot b_i$,

Proposition 1. *If* x *and* y *are primal and dual feasible solutions satisfying the conditions stated above then*

$$\sum_{j=1}^{n} c_j x_j \leq \alpha \cdot \beta \cdot \sum_{i=1}^{m} b_i y_i.$$

Proof.

$$\sum_{j=1}^{n} c_j x_j \leq \alpha \sum_{j=1}^{n} \left(\sum_{i=1}^{m} a_{ij} y_i \right) x_j = \alpha \sum_{i=1}^{m} \left(\sum_{j=1}^{n} a_{ij} x_j \right) y_i \leq \alpha \beta \sum_{i=1}^{m} b_i y_i. \quad (1)$$

The first and second inequalities follow from the primal and dual conditions respectively. The equality follows by simply changing the order of summation. □

The algorithm starts with a primal infeasible solution and a dual feasible solution; these are usually the trivial solutions $x = 0$ and $y = 0$. It iteratively improves the feasibility of the primal solution, and the optimality of the dual solution, ensuring that in the end a primal feasible solution is obtained and all conditions stated above, with a suitable choice of α and β, are satisfied. The primal solution is always extended integrally, thus ensuring that the final solution is integral. The improvements to the primal and the dual go hand-in-hand: the current primal solution is used to determine the improvement to the dual, and vice versa. Finally, the cost of the dual solution is used as a lower bound on OPT, and by Proposition 1, the approximation guarantee of the algorithm is $\alpha\beta$.

4   Primal-Dual Schema Applied to Set Cover

Problem 1. (Set cover) Given a universe U of n elements, a collection of subsets of U, $\mathcal{S} = \{S_1, \ldots, S_k\}$, and a cost function $c : \mathcal{S} \to \mathbf{Q}^+$, find a minimum cost sub-collection of $\mathcal{S}$ that covers all elements of U.

Define the *frequency* of an element to be the number of sets it is in. A useful parameter is the frequency of the most frequent element. Let us denote this by f. We obtain a factor f algorithm for the set cover problem using the primal–dual schema.

To formulate the set cover problem as an integer program, let us assign a variable x_S for each set $S \in \mathcal{S}$, which is allowed 0/1 values. This variable will be set to 1 iff set S is picked in the set cover. Clearly, the constraint is that for each element $e \in U$ we want that at least one of the sets containing it be picked.

$$
\begin{aligned}
\text{minimize} \quad & \sum_{S \in \mathcal{S}} c(S) x_S \\
\text{subject to} \quad & \sum_{S: e \in S} x_S \geq 1, \; e \in U \\
& x_S \in \{0, 1\}, \quad S \in \mathcal{S}
\end{aligned}
\quad (2)
$$

The LP-relaxation of this integer program is obtained by letting the domain of variables x_S be $1 \geq x_S \geq 0$. Since the upper bound on x_S is redundant, we get the following LP. A solution to this LP can be viewed as a fractional set cover.

$$\text{minimize} \quad \sum_{S \in \mathcal{S}} c(S) x_S$$

$$\text{subject to} \quad \sum_{S:\, e \in S} x_S \geq 1, \ e \in U \qquad\qquad (3)$$
$$x_S \geq 0, \qquad S \in \mathcal{S}$$

Introducing a variable y_e corresponding to each element $e \in U$, we obtain the dual program.

$$\text{maximize} \quad \sum_{e \in U} y_e$$

$$\text{subject to} \quad \sum_{e:\, e \in S} y_e levelc(S), \ S \in \mathcal{S} \qquad\qquad (4)$$
$$y_e \geq 0, \qquad e \in U$$

We will design the algorithm with $\alpha = 1$ and $\beta = f$. The complementary slackness conditions are:

Primal conditions:

$$\forall S \in \mathcal{S} : x_S \neq 0 \Rightarrow \sum_{e:\, e \in S} y_e = c(S).$$

Set S will be said to be *tight* if $\sum_{e:\, e \in S} y_e = c(S)$. Since we will increment the primal variables integrally, we can state the conditions as : *Pick only tight sets in the cover.*

Clearly, in order to maintain dual feasibility, we are not allowed to overpack any set.

Dual conditions:

$$\forall e : y_e \neq 0 \Rightarrow \sum_{S:\, e \in S} x_S \leq f$$

Since we will find a 0/1 solution for x, these conditions are equivalently to:
Each element having a nonzero dual value can be covered at most f times.
Since each element is in at most f sets, this condition is trivially satisfied for all elements.

The two sets of conditions naturally suggest the following algorithm:

Algorithm 1 (Set cover – factor f)

1. **Initialization:** $x \leftarrow 0;\ y \leftarrow 0$
2. Until all elements are covered, do:

    Pick an uncovered element, say e, and raise y_e until some set goes tight.
    Pick all tight sets in the cover and update x.
    Declare all the elements occurring in these sets as "covered".

3. Output the set cover x.

Theorem 2. *Algorithm 1 achieves an approximation factor of f.*

Proof. Clearly there will be no uncovered elements and no overpacked sets at the end of the algorithm. So, the primal and dual solutions will both be feasible. Since they satisfy the relaxed complementary slackness conditions with $\alpha = f$, by Proposition 1 the approximation factor is f.

Example 1. A tight example for this algorithm is provided by the following set system:

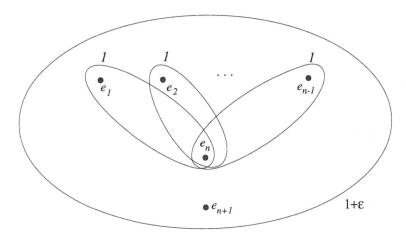

Here, $\mathcal{S}$ consists of $n - 1$ sets of cost 1, $\{e_1, e_n\}, \ldots, \{e_{n-1}, e_n\}$, and one set of cost $1 + \epsilon$, $\{e_1, \ldots, e_{n+1}\}$, for a small $\epsilon > 0$. Since e_n appears in all n sets, this set system has $f = n$.

Suppose the algorithm raises y_{e_n} in the first iteration. When y_{e_n} is raised to 1, all sets $\{e_i, e_n\}$, $i = 1, \ldots, n - 1$ go tight. They are all picked in the cover, thus covering the elements $e_1, \ldots, e_n$. In the second iteration, $y_{e_{n+1}}$ is raised to ϵ and the set $\{e_1, \ldots, e_{n+1}\}$ goes tight. The resulting set cover has a cost of $n + \epsilon$, whereas the optimum cover has cost $1 + \epsilon$.

5   A Combinatorial Algorithm

As stated in the Introduction, once an algorithm has been designed using the primal–dual schema, one can remove the scaffolding of linear programming and obtain a purely combinatorial algorithm. Let us illustrate this in the context of the weighted vertex cover problem, which is a special case of set cover with $f = 2$. For each vertex v, $t(v)$ is a initialized to its weight, and when $t(v)$ drops to 0, v is picked in the cover. $c(e)$ is the amount charged to edge e.

Algorithm 3

1. Initialization:
   $C \leftarrow \emptyset$
   $\forall v \in V, \ t(v) \leftarrow w(v)$
   $\forall e \in E, \ c(e) \leftarrow 0$
2. While C is not a vertex cover do:
   Pick an uncovered edge, say (u, v). Let $m = \min(t(u), t(v))$.
   $t(u) \leftarrow t(u) - m$
   $t(v) \leftarrow t(v) - m$
   $c(u, v) \leftarrow m$
   Include in C all vertices having $t(v) = 0$.
3. Output C.

One can show that this is a factor 2 approximation algorithm without invoking linear programs, by showing that the total amount charged to edges is a lower bound on OPT, and that the weight of cover C is at most twice the total amount charged to edges. Observe that the charges to the edges essentially correspond to the dual solution produced by Algorithm 1. It is easy to verify that this algorithm runs in linear, i.e., $O(|E|)$, time. This is currently the best known algorithm for the weighted vertex cover problem.

6   Discussion

It is instructive to compare the current status of primal–dual approximation algorithms with the (mature) status of exact primal–dual algorithms. In the latter setting, only one underlying mechanism is used: iteratively ensuring all complementary slackness conditions. On termination, an optimal (integral) solution to the LP is obtained. In the former setting, we are not seeking an optimal solution to the LP (since the LP may not have any optimal integral solutions), and so there is a need to introduce a further relaxation. Relaxing complementary slackness conditions (which itself can be carried out in more than one way) is only one of the possibilities (see [17] for an alternative mechanism). Another point of difference is that in the exact setting, more sophisticated dual growth algorithms have been given, e.g. [6]. In the approximation setting, other than [17], all primal–dual algorithms use a simple greedy dual growth algorithm.

So far, the primal–dual schema has been used for obtaining good integral solutions to an LP-relaxation. However, it seems powerful enough for the following more general scenario: when the **NP**-hard problem is captured not through an integer program, but in some other manner, and there is an LP that provides a relaxation of the problem. In this setting, the primal–dual schema will try to find solutions that are feasible for the original **NP**-hard problem, and are near-optimal in quality. Another possibility is to use the strong combinatorial conditions satisfied by solutions produced by the algorithms designed using the primal–dual. See [13] for one such use in cooperative game theory.

Some specific problems for which the primal–dual schema may be are right tool are:

Metric traveling salesman problem: The solution produced by the well known Christofedes' Algorithm [3] is within a factor of 3/2 of the subtour elemination LP-relaxation for this problem, [19]. However, the worst integrality gap example known is (essentially) 4/3. Can a 4/3 factor algorithm be obtained using this relaxation?

Steiner tree problem: The best approximation guarantee known is essentially 5/3, due to Promel and Steger [16], using structural properties established in [20]. A promising avenue for obtaining an improved guarantee is to use the bidirected cut LP-relaxation. This relaxation is exact for the minimum spanning tree problem. For the Steiner tree problem, the worst integrality gap known is (essentially) 8/7, due to M. Goemans. The best upper bound known on the integrality gap is 3/2 for quasi–bipartite graphs (graphs that do not contain edges connecting pairs of Steiner vertices), due to Rajagopalan and Vazirani [17]. Determine the integrality gap of this relaxation, and obtain an algorithm achieving this guarantee.

Steiner network problem: Jain gives a factor 2 algorithm for this problem [11]. However, it uses LP-rounding and has a prohibitive running time. Obtain a factor 2 combinatorial algorithm for this problem. A corollary of Jain's algorithm is that the integrality gap of LP-relaxation he used is bounded by 2. Therefore, this relaxation can be used as a lower bound for obtaining a factor 2 combinatorial algorithm.

References

[1] A. Agrawal, P. Klein, and R. Ravi. When trees colide: an approximation algorithm for the generalized Steiner network problem on networks. *SIAM J. Comp.*, 24:440–456, 1995.

[2] R. Bar-Yehuda and S. Even. A linear-time approximation algorithm for the weighted vertex cover problem. *J. Algorithms*, 2:198–203, 1981.

[3] N. Christofides. Worst-case analysis of a new heuristic for the travelling salesman problem. Technical report, Graduate School of Industrial Administration, Carnegie-Mellon University, Pittsburgh, PA, 1976.

[4] F. Chudak, T. Roughgarden, and D.P. Williamson. Approximate k-msts and k-steiner trees via the primal–dual method and lagrangean relaxation. In Manuscript, 2000.

[5] G.B. Dantzig, L.R. Ford, and D.R. Fulkerson. A primal–dual algorithm for linear programs. In H.W. Kuhn and A.W. Tucker, editors, *Linear inequalities and related systems*, pages 171–181, Princeton, NJ, 1956. Princeton University Press.

[6] J. Edmonds. Maximum matching and a polyhedron with 0,1-vertices. *J. Res. Nat. Bur. Standards*, B69:125–130, 1965.

[7] M.X. Goemans and D.P. Williamson. A general approximation technique for constrained forest problems. *SIAM J. Comp.*, 24:296–317, 1995.

[8] M.X. Goemans and D.P. Williamson. The primal–dual method for approximation algorithms and its applications to network design problems. pages 144–191, 1997.

[9] M. Grötschel, L. Lovász, and A. Schrijver. The ellipsoid method and its consequences in combinatorial optimization. *Combinatorica*, 1:169–197, 1981.

[10] M. Grötschel, L. Lovász, and A. Schrijver. *Geometric algorithms and combinatorial optimization*. Second edition. Springer-Verlag, Berlin, 1993.

[11] K. Jain. A factor 2 approximation algorithm for the generalized Steiner network problem. To appear in *Combinatorica*.

[12] K. Jain and V. V. Vazirani. Approximation algorithms for the metric facility location and k-median problems using the primal–dual schema and lagrangian relaxation. To appear in *J. ACM*.

[13] K. Jain and V. V. Vazirani. Applications of Approximation Algorithms to Cooperative Games Manuscript, 2000.

[14] H.W. Kuhn. The Hungarian method for the assignment problem. *Naval Research Logistics Quarterly*, 2:83–97, 1955.

[15] L. Lovász. *An algorithmic theory of numbers, graphs and convexity*. CBMS-NSF Regional Conference Series in Applied Mathematics, 50. SIAM, Philadelphia, PA, 1986.

[16] H. J. Prömel and A. Steger. RNC-approximation algorithms for the Steiner problem. In *Proc. Symposium on Theoretical Aspects of Computer Science*, volume 1200 of *Lecture Notes in Computer Science*, pages 559–570, Berlin, Germany, 1997. Springer-Verlag.

[17] S. Rajagopalan and V.V. Vazirani. On the bidirected cut relaxation for the metric Steiner tree problem. In *Proc. 10th ACM-SIAM Ann. Symp. Discret. Algorithms*, pages 742–751, 1999.

[Vaz01] V. V. Vazirani. *Approximation Algorithms*. Springer Verlag, Berlin, To appear, 2001.

[18] D.P. Williamson, M.X. Goemans, M. Mihail, and V.V. Vazirani. A primal–dual approximation algorithm for generalized Steiner network problems. *Combinatorica*, 15:435–454, 1995.

[19] L.A. Wolsey. Heuristic analysis, linear programming and branch and bound. *Mathematical Programming Study*, 13:121–134, 1980.

[20] A. Zelikovsky. An 11/6-approximation algorithm for the network Steiner problem. *Algorithmica*, 9:463–470, 1993.

Fourier Transforms and Quantum Computation

Umesh Vazirani[*]

University of California
Berkeley, CA 94720
USA
vazirani@cs.berkeley.edu

The foundations of computer science are built upon the modified Church-Turing thesis. This thesis states that any reasonable model of computation can be simulated by a probabilistic Turing Machine with at most polynomial factor simulation overhead (see [10] for a discussion). Early interest in quantum computation from a computer science perspective was sparked by results indicating that quantum computers violate the modified Church-Turing thesis [3,8]. The seminal work by Shor giving polynomial time algorithms for factorization and discrete logarithms [9] shook the foundations of modern cryptography, and gave a practical urgency to the area of quantum computation. All these quantum algorithms rely crucially upon properties of the Quantum Fourier transforms over finite Abelian groups. Indeed these properties are exactly what is required to solve a general problem known as the hidden subgroup problem, and it is easiest to approach the the algorithms for factoring and discrete logarithms as instances of this general approach. This survey paper focusses on presenting the essential ideas in a simple way, rather than getting the best results.

Can quantum computers solve NP-complete problems in polynomial time? This alluring prospect is unlikely to be resolved without a major breakthrough in complexity theory, since it has been proved [4] that there are oracles relative to which quantum computers do not provide a better than quadratic speed up for NP over exponential exhaustive search. The proof of this result relies on the hybrid argument, which we will present in this paper. There is a matching upper bound [5] showing that, in fact, this quadratic speed up can be achieved in general for any search problem.

This paper is based on course notes from my quantum computation course at Berkeley. The notes are accessible on the web at
  http:www.cs.berkeley.edu/~vazirani/quantum.html.

1   Qubits

The basic entity of quantum computation is a qubit (pronounced "cue-bit"), or a quantum bit. As an example, the electron in a hydrogen atom can be regarded as a two-level system: it can be in its ground state or in an excited state. Viewed as a classical system, the state of the electron could store a single bit of information:

---

[*] This research was supported by NSF Grant CCR-9800024, Darpa Grant F30602-00-2-0601.

ground $= 0$, excited $= 1$. But the general quantum state of the electron is a linear superposition of the two possibilities, and is called a qubit:

$$|\psi\rangle = \alpha|0\rangle + \beta|1\rangle \quad \alpha, \beta \in \mathbb{C} \quad \text{and} \quad |\alpha|^2 + |\beta|^2 = 1$$

In Dirac notation, used above, a column vector —called a "ket"— is denoted by $|\ \rangle$ — it has the advantage that the basis vectors can be labelled explicitly. (The $\{|0\rangle, |1\rangle\}$ basis is called the standard basis.) Thus the linear superposition written above is just a unit vector in a 2-dimensional complex vector space, and could alternately be written as a column vector $\begin{pmatrix} \alpha \\ \beta \end{pmatrix}$.

One may think of the linear superposition, $|\psi\rangle = \alpha|0\rangle + \beta|1\rangle$, as the electron's way of "not making up its mind" as to which of the 2 classical states it is in. This linear superposition is part of the private world of the electron, and for us to know the electron's state, we must make a measurement. A measurement in the $\{|0\rangle, |1\rangle\}$ basis yields $|0\rangle$ with probability $|\alpha|^2$, and $|1\rangle$ with probability $|\beta|^2$.

One important aspect of the measurement process is that it alters the state of the qubit: the new state is exactly the outcome of the measurement. i.e. if the outcome of the measurement of $|\psi\rangle = \alpha|0\rangle + \beta|1\rangle$ yields $|0\rangle$, then following the measurement, the qubit is in state $|0\rangle$. This implies that no further information about α, β can be collected by repeating the measurement.

The measurement above was made in the $|0\rangle, |1\rangle$ basis. More generally, we may choose any orthogonal basis $v, v^\perp$ and measure the qubit in that basis. To figure out the outcome, we just rewrite our state in that basis: $|\psi\rangle = \alpha'|v\rangle + \beta'|v^\perp\rangle$. Now the outcome is v with probability $|\alpha'|^2$, and $|v^\perp\rangle$ with probability $|\beta'|^2$. If the outcome of the measurement on $|\psi\rangle$ yields $|v\rangle$, then as before, the the qubit ends up in the state $|v\rangle$.

2   Two Qubits

How do we describe the quantum state of the two electrons in a pair of hydrogen atoms? Once again, if we were to view this as a classical system, then there are four possibilities: $\{00, 01, 10, 11\}$, which can represent 2 bits of classical information. Now the general quantum state of this system is a linear superposition of these four states:

$$|\psi\rangle = \alpha_{00}|00\rangle + \alpha_{01}|01\rangle + \alpha_{10}|10\rangle + \alpha_{11}|11\rangle$$

where by $|ab\rangle$ we mean the first qubit has state a, and second qubit has state b. This is just Dirac notation for the unit vector in $\mathbb{C}^4$:

$$\begin{pmatrix} \alpha_{00} \\ \alpha_{01} \\ \alpha_{10} \\ \alpha_{11} \end{pmatrix}$$

where $\alpha_{jk} \in C, \sum |\alpha_{jk}|^2 = 1$.

If we were to measure the two qubits, then the probability that the first qubit is in state j, and the second qubit is in state k is $P(j,k) = |\alpha_{jk}|^2$. Following the measurement, the state of the two qubits is $|\psi\rangle = |jk\rangle$. What happens if we measure just the first qubit? The probability that the first qubit is 0 is the same as in the case that we measure both qubits: $\Pr\{\text{1st bit} = 0\} = |\alpha_{00}|^2 + |\alpha_{01}|^2$ But the new state of the system consists of those parts of the superposition that are consistent with the outcome of the measurement– but normalized to be a unit vector:

$$|\phi\rangle = \frac{(\alpha_{00}|01\rangle + \alpha_{01}|00\rangle)}{\sqrt{|\alpha_{00}|^2 + |\alpha_{01}|^2}}$$

Example:
Consider the following state, which is called an EPR pair or a Bell state:

$$|\Psi^-\rangle = \frac{1}{\sqrt{2}}(|01\rangle - |10\rangle)$$

Measuring either bit of $|\Psi^-\rangle$ in the standard basis yields a 0 with probability $1/2$, and 1 with probability $1/2$. However, if the outcome of measuring the first bit is a 0, then the second bit is necessarily a 1, and vice versa.

Moreover, the state $|\Psi^-\rangle$ is rotationally symmetric, in the sense that if we measure the two qubits in an arbitrary basis — $v, v^\perp$ — the outcomes for the two qubits will be opposite. This follows from the fact that $|\Psi^-\rangle = \frac{1}{\sqrt{2}}(|vv^\perp\rangle - |v^\perp v\rangle)$.

In 1964, John Bell showed that these properties of an EPR state have verifiable consequences that distinguish quantum mechanics from any hidden variable theory. Consider an experiment where two particles are initialized in an EPR state, and then are measured at two distant locations A and B. At each location, one of two measurements is chosen at random, according to random inputs X_A and X_B respectively. The challenge for a hidden variable theory is to achieve the same probability distribution on outcomes of the measurements, in the following abstract setting: the initialization of the two particles is now a classical probabilistic process, and can be abstractly modeled as assigning the same arbitrarily long random string r (the hidden variable) to each particle. The outcome of the experiment at location A is allowed to be an arbitrary function of X_A and the hidden variable r (similarly for location B).

The following concrete communication problem may be stated in this setting: Let A and B be two parties who share a common random string r. They receive as input random bits X_A, X_B, and must output bits a, b respectively, such that $X_A \wedge X_B = a \oplus b$. How small can they make their probability of failure? It is easy to show that the optimum strategy for A and B is deterministic, and that A and B must fail with probability at least $1/4$.

In the quantum setting, A and B now share an EPR pair $|\Psi^-\rangle$. As before, they receive bits X_A, X_B, and try to output bits a, b respectively, such that $a, b \in 0, 1$, s.t. $X_A \wedge X_B = a \oplus b$. Surprisingly, there is a strategy for A and B such that they fail with probability $\frac{3-\sqrt{2}}{8} < 1/4$:

- if $X_A = 0$, then A measures (her half of $|\Psi^-\rangle$) in the standard basis, and outputs the result.
- if $X_A = 1$, then A measures in the basis rotated by $\pi/8$, and outputs the result.
- if $X_B = 0$, then B measures in the standard basis, and outputs the complement of the result.
- if $X_B = 1$, then B measures in the basis rotated by $-\pi/8$, and outputs the complement of the result.

To analyze this strategy, first notice that if A and B share an EPR pair, and A measures her qubit in the standard basis, and B measures his qubit in a basis rotated by θ, then the two outcomes are opposite with probability $\cos^2(\theta)$, and equal with probability $\sin^2(\theta)$. Adding the failure probability in the four cases now gives: the total probability that A and B fail $= 1/4(0 + \sin^2(\pi/8) + \sin^2(\pi/8) + 1/2) = \frac{3-\sqrt{2}}{8} \equiv .2 < 1/4$. This shows that no hidden variable theory is consistent with the predictions of quantum physics.

3   Tensor Products

The state of a single quantum bit is a unit vector in the Hilbert space $\mathbb{C}^2$. The state of two quantum bits is a unit vector in the tensor product of this space with itself $\mathbb{C}^2 \otimes \mathbb{C}^2 = \mathbb{C}^4$. Using Dirac "ket" notation, we write the basis of $\mathbb{C}^2 \otimes \mathbb{C}^2$ as

$$\{|0\rangle \otimes |0\rangle, |0\rangle \otimes |1\rangle, |1\rangle \otimes |0\rangle, |1\rangle \otimes |1\rangle\}$$

We will often write $|0\rangle \otimes |0\rangle$ as $|0\rangle|0\rangle$ or $|00\rangle$.

In general, we represent an n-particle system by n copies of $\mathbb{C}^2$ tensored together. We will often write $(\mathbb{C}^2)^{\otimes n} = \mathbb{C}^{2^n}$. So the state of an n-qubit system can be written as

$$|\psi\rangle = \sum_{x \in \{0,1\}^n} \alpha_x |x\rangle \qquad\qquad \sum_x |\alpha_x|^2 = 1$$

This means that nature must 'remember' 2^n complex numbers just to keep track of the state of an n-particle system. The idea behind quantum computation is to harness the ability of nature to manipulate these exponentially many α_xs.

4   Quantum Gates and Quantum Circuits

Quantum physics requires that the evolution of a quantum state in time is given by a unitary transformation. Such a transformation U preserves inner product of the underlying Hilbert space.

A unitary transformation on n qubits is $c - local$ if it operates nontrivially on at most c of the qubits, and preserves the remaining qubits. Quantum computation may be thought of as the study of those unitary transformations that can be realized as a sequence of $c - local$ unitary transformations. i.e.

$$\mathcal{U} = \mathcal{U}_1 \mathcal{U}_2 \cdots \mathcal{U}_k,$$

where the $\mathcal{U}_i$ are $c - local$ and k is bounded by a polynomial in n.

Indeed, it has been shown that we can simplify the picture further, and restrict attention to certain special $2 - local$ unitary transformations, also called elementary quantum gates [2]:

Rotation

A rotation or phase shift through an angle θ can be represented by the matrix

$$\mathcal{U} = \begin{pmatrix} \cos\theta & -\sin\theta \\ \sin\theta & \cos\theta \end{pmatrix}.$$

This can be thought of as rotation of the axes (see Figure 1).

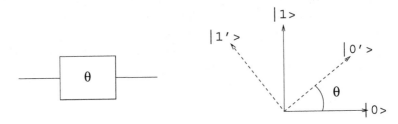

Fig. 1. Rotation.

Hadamard Transform

If we reflect the axes in the line $\theta = \pi/8$ we get the *Hadamard transform* (see Figure 2). This can be represented by the matrix

$$\mathcal{H} = \frac{1}{\sqrt{2}} \begin{pmatrix} 1 & 1 \\ 1 & -1 \end{pmatrix}.$$

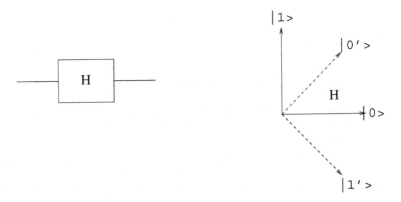

Fig. 2. Hadamard transform.

Controlled NOT

The controlled NOT gate (Figure 3) operates on 2 qubits and can be represented by the matrix

$$\begin{pmatrix} 1 & 0 & 0 & 0 \\ 0 & 1 & 0 & 0 \\ 0 & 0 & 0 & 1 \\ 0 & 0 & 1 & 0 \end{pmatrix},$$

where the basis elements are (in order) $|00\rangle$, $|01\rangle$, $|10\rangle$, $|11\rangle$. If inputs a and b are basis states, then the outputs are a and $a \oplus b$.

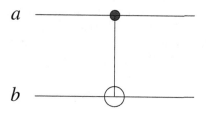

Fig. 3. Controlled NOT.

Any unitary transformation can be approximated by using just rotations, the Hadamard transform, and controlled NOT gates [2].

5   Primitives for Quantum Computation

Let us consider how we can simulate a classical circuit with a quantum circuit. The first observation is that quantum evolution is unitary, and therefore reversible (the effect of a unitary transformation U can be undone by applying its adjoint). Therefore if the classical circuit computes the function $f : \{0,1\}^n \to \{0,1\}$ then we must allow the quantum circuit to output $|x\rangle\,|f(x)\rangle$ on input x. It follows from the work of Bennett [1] on reversible classical computation that, if $C(f)$ be the size of the smallest classical circuit that computes f, then there exists a quantum circuit of size $O(C(f))$ which, for each input x to f, computes the following unitary transformation U_f on m qubits:

$$U_f : |x\rangle\,|y\rangle \to |x\rangle\,|y \oplus f(x)\rangle$$

In general, though, we don't need to feed U_f a classical state $|x\rangle$. If we feed U_f a superposition

$$\sum_{x \in \{0,1\}^n} \alpha_x\,|x\rangle\,|0\rangle$$

then, by linearity,

$$U_f\left(\sum_{x\in\{0,1\}^n}\alpha_x\,|x\rangle\,|0\rangle\right)=\sum_{x\in\{0,1\}^n}\alpha_x U_f\left(|x\rangle\,|0\rangle\right)=\sum_{x\in\{0,1\}^n}\alpha_x\,|x\rangle\,|f\left(x\right)\rangle.$$

At first sight it might seem that we've computed $f\left(x\right)$ simultaneously for each basis state $|x\rangle$ in the superposition. However, were we to make a measurement, we would observe $f\left(x\right)$ for only one value of x.

Hadamard Transform

The second primitive is the Hadamard transform H_{2^n}, which corresponds to the Fourier transform over the abelian group Z_2^n. The correspondence is based on identifying $\{0,1\}^n$ with Z_2^n, where the group operation is bitwise addition modulo 2. One way to define H_{2^n} is as the $2^n\times 2^n$ matrix in which the (x,y) entry is $2^{-n/2}\left(-1\right)^{x\cdot y}$. An equivalent way is as follows: let H be the unitary transformation on one qubit defined by the matrix

$$\begin{pmatrix}\frac{1}{\sqrt{2}} & \frac{1}{\sqrt{2}}\\ \frac{1}{\sqrt{2}} & -\frac{1}{\sqrt{2}}\end{pmatrix}.$$

When we combine quantum circuits that operate on disjoint qubits, the matrix for the combined circuit is formed by the tensor product of the matrices for the individual circuits. Thus $H_{2^n}=H^{\otimes n}$, or H tensored with itself n times.

Applying the Hadamard transform (or the Fourier transform over Z_2^n) to the state of all zeros gives an equal superposition over all 2^n states

$$\mathcal{H}_{2^n}|0\cdots 0\rangle=\frac{1}{\sqrt{2^n}}\sum_{x\in\{0,1\}^n}|x\rangle.$$

Applying the Hadamard transform to the computational basis state $|u\rangle$ modifies the above superposition, by a phase:

$$\mathcal{H}_{2^n}|u\rangle=\frac{1}{\sqrt{2^n}}\sum_{x\in\{0,1\}^n}(-1)^{u\cdot x}|x\rangle.$$

Fourier Sampling

In general, if we start with a state $|\phi\rangle=\sum_x\alpha_x|x\rangle$, after applying the Fourier transform over Z_2^n, we obtain the new state $|\hat{\phi}\rangle=\sum_x\hat{\alpha}_x|x\rangle$. Notice that this transform can be computed by applying only n single qubit gates, whereas it is computing the Fourier transform on a 2^n dimensional vector. However, the output of the Fourier transform is not accessible to us. To read out the answer, we must make a measurement, and now we obtain x with probability $|\hat{\alpha}_x|^2$. This process of computing the Fourier transform and then performing a measurement is called Fourier sampling, and is one of the basic primitives in quantum computation (see [3] for further discussion about the complexity of this primitive).

In the next section, we shall consider Fourier transforms over arbitrary finite abelian groups. Kitaev [7] showed that there are efficient quantum circuits for computing the quantum Fourier transform over an arbitrary finite abelian group, and therefore for Fourier sampling over an arbitrary finite abelian group. There is also work by Hales and Hallgren [6] that obtains this result by a completely different approach.

6   Fourier Transforms over Finite Abelian Groups

Let G be a finite abelian group. The characters of G are homomorphisms $\chi_j : G \to \mathbb{C}$. There are exactly $|G|$ characters, and they form a group, called the dual group, and denoted by $\hat{G}$. The Fourier transform over the group G is given by:

$$|g\rangle \mapsto \frac{1}{\sqrt{|G|}} \sum_j \chi_j(g)|j\rangle$$

Consider, for example $G = Z_N$. The characters are defined by $\chi_j(1) = \omega^j$ and $\chi_j(k) = \omega^{jk}$. And the Fourier transform is given by the familiar matrix F, with $F_{j,k} = \frac{1}{\sqrt{N}}\omega^{jk}$.

In general, let $G \cong \mathbb{Z}_{N_1} \times \mathbb{Z}_{N_2} \times \cdots \times \mathbb{Z}_{N_l}$, so that any $g \in G$ can be written equivalently as $(a_1, a_2, \ldots, a_l)$, where $a_i \in \mathbb{Z}_{N_i}$. Now, for each choice of $k_1, \ldots, k_l$ we have a character given by the mapping:

$$\chi_{k_1,\ldots,k_l}(a_1, a_2, \ldots, a_l) = \omega_{N_1}^{k_1 a_1} \cdot \omega_{N_2}^{k_2 a_2} \cdot \ldots \cdot \omega_{N_l}^{k_l a_l}$$

Finally, the Fourier transform of $(a_1, a_2, \ldots, a_l)$ can be defined as

$$(a_1, a_2, \ldots, a_l) \mapsto \frac{1}{\sqrt{|G|}} \sum_{(k_1,\ldots,k_l)} \omega_{N_1}^{k_1 a_1} \omega_{N_2}^{k_2 a_2} \cdot \ldots \cdot \omega_{N_l}^{k_l a_l} |k_1 \cdots k_l\rangle$$

7   Subgroups and Cosets

Corresponding to each subgroup $H \subseteq G$, there is a subgroup $H^\perp \subseteq \hat{G}$, defined as $H^\perp = \{k \in \hat{G} \mid k(h) = 1 \ \forall h \in H\}$, where $\hat{G}$ is the dual group of G. $|H^\perp| = \frac{|G|}{|H|}$. The Fourier transform over G maps an equal superposition on H to an equal superposition over $H^\perp$:

Claim

$$\frac{1}{\sqrt{|H|}} \sum |h\rangle \stackrel{FT_G}{\mapsto} \sqrt{\frac{|H|}{|G|}} \sum_{k \in H^\perp} |k\rangle$$

Proof. The amplitude of each element $k \in H^\perp$ is $\frac{1}{\sqrt{|G|}\sqrt{|H|}} \sum_{h \in H} k(h) = \frac{\sqrt{|H|}}{\sqrt{|G|}}$.

But since $|H^\perp| = \frac{|G|}{|H|}$, the sum of squares of these amplitudes is 1, and therefore the amplitudes of elements not in $H^\perp$ is 0.

The Fourier transform over G treats equal superpositions over cosets of H almost as well:

Claim

$$\frac{1}{\sqrt{|H|}} \sum_{h \in H} |hg\rangle \overset{FT_G}{\mapsto} \sqrt{\frac{|H|}{|G|}} \sum_{k \in H^\perp} \chi_g(k)|k\rangle$$

Proof. This follows from the convolution-multiplication property of Fourier transforms. An equal superposition on the coset Hg can be obtained by convolving the equal superposition over the subgroup H with a delta function at g. So after a Fourier transform, we get the pointwise multiplication of the two Fourier transforms: namely, an equal superposition over $H^\perp$, and χ_g.

Since the phase $\chi_g(k)$ has no effect on the probability of measuring $|k\rangle$, Fourier sampling on an equal superposition on a coset of H will yield a uniformly random element $k \in H^\perp$. This is a fundamental primitive in the quantum algorithm for the hidden subgroup problem.

Claim. Fourier sampling performed on $|\Phi\rangle = \frac{1}{\sqrt{|H|}} \sum_{h \in H} |hg\rangle$ gives a uniformly random element $k \in H^\perp$.

8   The Hidden Subgroup Problem

Let G again be a finite abelian group, and $H \subseteq G$ be a subgroup of G. Given a function $f : G \to S$ which is constant on cosets of H and distinct on distinct cosets (i.e. $f(g) = f(g')$ iff there is an $h \in H$ such that $g = hg'$), the challenge is to find H.

The quantum algorithm to solve this problem is a distillation of the algorithms of Simon and Shor. It works in two stages:

Stage I. Setting Up a Random Coset State:

Start with two quantum registers, each large enough to store an element of the group G. Initialize each of the two registers to $|0\rangle$. Now compute the Fourier transform of the first register, and then store in the second register the result of applying f to the first register. Finally, measure the contents of the second register. The state of the first register is now a uniform superposition over a random coset of the hidden subgroup H:

$$|0\rangle |0\rangle \quad \overset{FT_G \otimes I}{\longrightarrow} \quad \frac{1}{\sqrt{|G|}} \sum_{a \in G} |a\rangle |0\rangle$$

$$\overset{f}{\longrightarrow} \quad \frac{1}{\sqrt{|G|}} \sum_{a \in G} |a\rangle |f(a)\rangle$$

$$\overset{\text{measure 2nd reg}}{\longrightarrow} \quad \frac{1}{\sqrt{|H|}} \sum_{h \in H} |hg\rangle$$

Stage II. Fourier Sampling:

Compute the Fourier transform of the first register and measure. By the last claim of the previous section, this results in a random element of $H^\perp$. i.e. random $k : k(h) = 0 \; \forall h \in H$. By repeating this process, we can get a number of such random constraints on H, which can then be solved to obtain H.

Example. Simon's Algorithm: In this case $G = Z_2^n$, and $H = \{0, s\}$. Stage I sets up a random coset state $1/\sqrt{2}|x\rangle + 1/\sqrt{2}|x + s\rangle$. Fourier sampling in stage II gives a random $k \in Z_2^n$ such that $k \cdot s = 0$. Repeating this $n - 1$ times gives

$n - 1$ random linear constraints on s. With probability at least $1/e$ these linear constraints have full rank, and therefore s is the unique non-zero solution to these simultaneous linear constraints.

9   Factoring and Discrete Log

Factoring is a problem of great practical interest, and its classical intractability forms the basis of the RSA cryptosystem. The factoring problem may be stated as given N find $N_1, N_2 > 1$ such that $N = N_1 N_2$.

A closely related problem to factoring is *order finding*. To define this problem, recall that:

The set of integers that are relatively prime to N form a group under the operation of multiplication modulo N: $Z_N^* = \{x \in Z_N : gcd(x, N) = 1\}$.

Let $x \in Z_N^*$. The order of x (denoted by $ord_N(x)$) is $min_{r \geq 1} x^r \equiv 1 \bmod N$.

Lemma. Let N be an odd composite (not a prime power), and let $x \bmod N$ be random and relatively prime to N. Let r be the order of $x \bmod N$. Then with probability at least $1/2$, r is even and $x^{r/2} \neq \pm 1 \bmod N$.

The proof of this lemma is simple and relies on the Chinese remainder theorem. The lemma says that if we know how to compute the order of a random $x \in Z_N^*$, then with high probability we can find a non-trivial square root of $1 \bmod N$, and therefore factor N. To see this let $a = x^{r/2}$. Then a is a non-trivial square root of $1 \bmod N$, since $a^2 = 1 \bmod N$, but $a \neq \pm 1 \bmod N$. Therefore $N | (a+1)(a-1)$, but $N \nmid (a+1)$ and $N \nmid (a-1)$. Thus $gcd(N, a+1)$ must be a proper factor of N, and can be efficiently found via Euclid's algorithm.

Example. Suppose that $N = 15$. Then $4^2 \equiv 1 \bmod N$ while $4 \neq \pm 1 \bmod N$ hence $gcd(15, 4-1) = 5$ and $gcd(15, 4+1) = 3$ are both nontrivial factors of 15.

So now the task of factoring N is reduced to the task of computing the order of a given $x \in Z_N^*$. Recall that $|Z_N^*| = \Phi(N)$, where $\Phi(N)$ is the Euler Phi function. If $N = p_1^{e_1} \cdots p_k^{e_k}$ then $\phi(N) = (p_1 - 1)p_1^{e_1 - 1} \cdots (p_k - 1)p_k^{e_k - 1}$. Clearly, $ord_N(x) | \Phi(N)$.

Consider the function $f : Z_{\Phi(N)} \to Z_N$, where $f(a) = x^a \bmod N$. Then $f(a) = 1$ if $a \in \langle r \rangle$, where $r = ord_N(x)$, and $\langle r \rangle$ denotes the subgroup of Z_N^* generated by r. Similarly if $a \in \langle r \rangle + k$, a coset of $\langle r \rangle$, then $f(a) = x^k \bmod N$. Thus f is constant on cosets of $H = \langle r \rangle$.

The quantum algorithm for finding the order r or x first uses f to set up a random coset state, and then does Fourier sampling to obtain a random element from $H^\perp$. Notice that the random element will have the form

$$k = s \cdot \frac{\phi(N)}{r}$$

where s is picked randomly from $\{0, \dots, r-1\}$. If $gcd(s, r) = 1$ (which holds for random s with reasonably high probability), $gcd(k, \phi(N)) = \phi(N)/r$. From this it is easy to recover r. There is no problem discarding bad runs of the algorithm, since the correct value of r can be used to split N into non-trivial factors.

Discrete Log Problem:

Computing discrete logarithms is another fundamental problem in modern cryptography. Its assumed hardness underlies the Diffie-Helman cryptosystem.

In the Discrete Log problem is the following: given a prime p, a generator g of Z_p^* (Z_p^* is cyclic if p is a prime), and an element $x \in Z_p^*$; find r such that $g^r \equiv x \bmod p$.

Define $f : Z_{p-1} \times Z_{p-1} \to Z_p^*$ as follows: $f(a,b) = g^a x^{-b} \bmod p$.

Notice that $f(a,b) = 1$ exactly when $a = br$. Equivalently, when $(a,b) \in \langle (r,1) \rangle$, where $\langle (r,1) \rangle$ denotes the subgroup of $Z_{p-1} \times Z_{p-1}$ generated by $(r,1)$.

Similarly, $f(a,b) = g^k$ for $(a,b) \in \langle (r,1) \rangle + (k,0)$. Therefore, f is constant on cosets of $H = \langle (r,1) \rangle$.

Again the quantum algorithm first uses f to set up a random coset state, and then does Fourier sampling to obtain a random element from $H^{\perp}$. i.e. (c,d) such that $rc + d = 0 \bmod p - 1$. For a random such choice of (c,d), with reasonably high probability $gcd(c, p-1) = 1$, and therefore $r = -dc^{-1} \bmod p - 1$. Once again, it is easy to check whether we have a good run, by simply computing $g^r \bmod p$ and checking to see whether it is equal to x.

Making It All Work

To make the simplified picture of the factoring and discrete log problem rigorous, we have to do a little more work. In the case of the discrete log problem, the main remaining issue is computing the Fourier transform over the group $Z_{p-1} \times Z_{p-1}$. This is not straightforward, since the Fourier transform over this group does not have a straightforward tensor decomposition. There are two main approaches: one is to use Kitaev's phase estimation technique [7], to compute Fourier transforms for any abelian group. A second technique, clean Fourier sampling [6], shows that Fourier transforms are robust under changes in the underlying group. Thus if we wish to Fourier sample the state $|\phi\rangle$ with respect to the group Z_N, the result can be closely approximated by interpreting $|\phi\rangle$ as a state over Z_M for $M >> N$, and Fourier sampling over Z_M and considering the conditional distribution on N evenly spaced points $\lceil \frac{jM}{N} \rceil$. Now if we let $M = 2^k$, computing the quantum Fourier transform over Z_M is easy.

In the case of factoring, the problem is more complicated. The issue is that knowing the Euler Phi function $\Phi(N)$ is as hard as factoring, since there is a polynomial time classical algorithm to factor N given $\Phi(N)$. The technique of clean Fourier sampling still works though, and we can do the Fourier sampling with respect to a large enough Z_M. The key observation is that the algorithm does not need the measured value k, but only the ratio $\frac{k}{\Phi(N)}$, which is closely approximated by the ratio $\frac{k'}{M}$ in the clean Fourier sampling.

10   Quantum Algorithms for NP?

Consider the satisfiability (SAT) problem: given a Boolean formula $f(x_1, x_2, \dots, x_n)$, is there an boolean assignment to $x_1, \dots, x_n$ that satisfies f? Brute force search would take $O(N)$ steps, where $N = 2^n$. Is there a quantum algorithm that can solve this problem in polynomial in n steps? We will abstract this problem in a black box or oracle model as follows: assume that the input to the problem is a table with N boolean entries, and the task is to find whether any entry in the table is 1. Classical algorithms can have random access to the table entries, and quantum algorithms can query the entries of the table in superposition. We will show, using the hybrid argument [4], that any quantum algorithm for this problem must make $\Omega(\sqrt{N})$ queries to the table.

Consider any quantum algorithm A for solving the search problem. First do a test run of A on function $f \equiv 0$. Define the query magnitude of x to be $\sum_t |\alpha_{x,t}|^2$, where $\alpha_{x,t}$ is the amplitude with which A queries x at time t. The expected sum of the query magnitudes $E_x \left(\sum_t |\alpha_{x,t}|^2 \right) = T/N$, where T is the total number of queries made by A. Thus $min_x \left(\sum_t |\alpha_{x,t}|^2 \right) \leq T/N$. For any such x, by the Cauchy-Schwarz inequality, $\sum_t |\alpha_{x,t}| \leq T/\sqrt{N}$.

Let $|\phi_0\rangle, \ldots, |\phi_T\rangle$ be the states of A_f during its run on f. Now run algorithm A on the function $g : g(x) = 1, g(y) = 0$ for $y \neq x$. Suppose the final state of A_g is $|\psi_T\rangle$. We will show that $|||\phi_T\rangle - |\psi_T\rangle||$ must be small.

Claim. $|\psi_T\rangle = |\phi_T\rangle + |E_0\rangle + \ldots |E_{T-1}\rangle$, where $|||E_t\rangle|| \leq \alpha_{x,t}$.

Proof. Consider two runs of the algorithm A, which differ only in the tth step: on the tth step the first run queries f and the second queries g. Both runs query f on the first $t-1$ steps, and both query g starting with the $t+1^{st}$ step. Then at the end of the tth step, the state of the first run is $|\phi_t\rangle$, whereas the state of the second run is $|\phi_t\rangle + |F_t\rangle$, where $|||F_t\rangle|| \leq |\alpha_{x,t}|$. Now, if U is the unitary transformation describing the remaining $T-t$ steps, then the final state after T steps for the two runs are $U|\phi_t\rangle$ and $U(|\phi_t\rangle + |F_t\rangle)$ respectively. The latter can be written as $U|\phi_t\rangle + |E_t\rangle$, where $|E_t\rangle = U|F_t\rangle$. The claim now follows by 'walking from' the pure f run to the pure g run by switching one additional query in each step, thus showing that the total error vector is the sum of the $|E_i\rangle$'s.

It follows from the above claim that $|||\phi_T\rangle - |\psi_T\rangle| \leq \sum_t |\alpha_{x,t}| \leq T/\sqrt{N}$. It was proved in [3] that if $|||\phi\rangle - |\psi\rangle|| \leq \epsilon$ then the probability distributions that result from measuring $|\phi\rangle$ and $|\psi\rangle$ are with in a total variation distance of 4ϵ of each other. Therefore, for the algorithm A to distinguish f from g with constant probability, it follows that $T/\sqrt{N} = \Omega(1)$. i.e. $T = \Omega(\sqrt{(N)})$.

There is a matching upper bound due to Grover [5], that gives a quantum algorithm for search that runs in $O(\sqrt{N})$ steps.

Acknowledgements. This paper is based on course notes from my course on quantum computation at Berkeley. I wish to thank the students who scribed the lectures, especially Allison Coates, Lawrence Ip, Kunal Talwar, Lisa Hales and Ziv Bar-Yassef.

References

[1] Bennett, C. H., "Logical reversibility of computation," *IBM J. Res. Develop.*, Vol. 17, 1973, pp. 525-532.

[2] Barenco, A., Bennett, C., Cleve, R., DiVincenzo, D., Margolus, N., Shor, P., Sleator, T., Smolin, J., and Weinfurter, H., "Elementary gates for quantum computation," Phys. Rev. A **52**, 3457 (1995).

[3] Bernstein E and Vazirani U, 1993, Quantum complexity theory, *SIAM Journal of Computation* **26** 5 pp 1411–1473 October, 1997.

[4] Bennett, C.H., Bernstein, E., Brassard, G. and Vazirani, U., "Strengths and weaknesses of quantum computation," SIAM J. Computing, 26, pp. 1510-1523 (1997).

[5] Grover, L., "Quantum mechanics helps in searching for a needle in a haystack,' ' Phys. Rev. Letters, 78, pp. 325-328 (1997).

[6] L. Hales and S. Hallgren. Quantum Fourier Sampling Simplified. In *Proceedings of the Thirty-first Annual ACM Symposium on the Theory of Computing*, pages 330-338, Atlanta, Georgia, 1-4 May 1999.

[7] Alexei Kitaev. Quantum measurements and the abelian stabilizer problem. ECCC Report TR96-003, 1996.

[8] D. Simon. "On the power of quantum computation." In *Proc. 35th Symposium on Foundations of Computer Science (FOCS), 1994.*

[9] Shor P W, Polynomial-time algorithms for prime factorization and discrete logarithms on a quantum computer, *SIAM J. Comp.*, **26**, No. 5, pp 1484–1509, October 1997.

[10] Vazirani, U., "On the power of quantum computation," *Philosophical Transactions of the Royal Society of London, Series A*, 356:1759-1768, August 1998.

Author Index

Lecture Notes in Computer Science

For information about Vols. 1–2201
please contact your bookseller or Springer-Verlag